A DRUNK WITH A PASSPORT

BY
B.O.L.O. LEE

Copyright © 2026 by Kao B.O.L.O. Lee

All rights reserved. No portion of this book may be used or reproduced, stored in a retrieval system or transmitted in any form by any means – electronic, mechanical, photocopy, recording, or otherwise – without the written permission of the publisher except in the case of brief quotations embodied in printed reviews.

ISBN: 9798243079471

For Victoria and Zoey.

CONTENTS

Prologue	7
Chapter 1: One Mountain a Day	9
Chapter 2: Rock Bottom	14
Chapter 3: One Way Ticket	18
Chapter 4: Bangkok	21
Chapter 5: Round Two	26
Chapter 6: Koh Samui	32
Chapter 7: The Three Amigos	44
Chapter 8: Koh Phangan	49
Chapter 9: Full Moon Party	59
Chapter 10: The Long Way Home	68
Chapter 11: Back to the Grind	74
Chapter 12: The Dating Comedy Show	80
Chapter 13: Baptism	88
Chapter 14: Island Hopping	96
Chapter 15: The Durian Farm	99
Chapter 16: Koh Yao Yai	104
Chapter 17: Phuket	115
Chapter 18: Hell On Earth	119
Chapter 19: Red Hot Knives	124

Chapter 20: The Promise	131
Chapter 21: Minneapolis	135
Chapter 22: Sacramento	139
Chapter 23: Denver	144
Chapter 24: Seoul	152
Chapter 25: Manila	158
Chapter 26: Coron	165
Chapter 27: Brick to the Face	175
Chapter 28: Leaving Paradise	178
Epilogue	182

Prologue:

"Wake up, Son, we have to go right now. The Vietnamese are coming!"

That was my earliest memory.

The night was dark and loud—I could feel something was wrong and see panic shuffling like ghosts through the room. Uncles, aunts, cousins, shadows stuffing things into other things. She grabbed my hand and we started running.

I didn't have shoes. Not back then. The ground was cold and wet, leaves rustling, rocks and sticks poking, small brown nuts crunching under my toes. I didn't know what they were, only that they cut and hurt and shined faintly in the wet moonlight. Mom had my baby brother strapped to her chest and a basket of rice, clothing, and whatever she could salvage tied to her back. Walking when we could but mostly sliding up and down steep muddy hills, in the jungle, in the dark. Crawling. Clutching. Bleeding.

We marched in a single line, silently, except for the sounds of feet, fear, and cries, through mud and darkness. Every step had to stay on the trail. The world beyond a wrong footstep was a landmine wailing.

Years later I finally told Mom about the nightmare that still haunts, the smell of rain, the mysterious nuts underfoot, the sound of her breathing hard. She cried. Said those were water chestnuts. Said I was only three years old. Said it was the longest night of her life. Dad had gone ahead with my grandparents. He was supposed to come back for us. He never made it in time. The Vietnamese came early, so Mom did what mothers do, she carried everything and everyone to safety.

When my feet gave out, she carried me too. She said we fell often - mud, tears, blood, hunger, everything mixing into one long night. "Your brother cried, you cried, I cried," she tells me now. "But we kept going."

I always ask her, "How did you do it, Mom?"

And she always says, "One mountain a day, Son."

It took me half a life to understand that it wasn't about geography. It was philosophy. It was survival disguised as mom talk.

She said, "If you look at all the mountains, you'll give up before you start. But if you just climb the one in front of you, you'll make it. Then tomorrow, another one. That's how you live, Son. That's how you survive. One mountain a day".

Chapter 1: One Mountain a Day

Was I always a drunk?

No.

There was a time when I woke up every morning at four a.m. like a damn spring chicken—full of useless energy only because I didn't drink myself into oblivion the night before. Back then, I hated alcohol. Beer tasted like cold piss. Shots burned straight through the soul. And if I absolutely had to drink, it was Jack drowned in an entire can of Coke so I couldn't taste the poison.

I'd make my bed immediately, then scribble my daily goals down before making coffee—unless I was on one of my 90-day caffeine breaks to "reset my tolerance," in which case I'd choke down green tea instead. After that, I'd hit the gym and post my morning routine on Instagram like I was some kind of motivational guru with approximately 100 followers. Most of those followers were my little cousins, and they ate it up.

Life was gentle with me.

A golden stretch of years when I was the good son, the good husband, the good corporate soldier. A sensible man in a sensible button-up shirt, sending a hundred polite emails a day to the same tired sales reps— translating Corporate into Human so they could do the real work while I pretended my leadership mattered.

And I mattered. A lot. To me.

I had a wife who still let me touch her. Two beautiful baby boys—three and five—those chunky cheeks still ambushing me in Facebook Memories, hitting me with joy and grief in the same heartbeat. A nice house. Two cars. The American Dream, laminated and framed. And I was good at it too—the best sales manager in the entire nation for my company.

Then I wrote and published my first book.

I was proud of it in the raw, stupid way a refugee kid is proud when he does the impossible. Me—the kid who didn't speak a word of English until nearly ten. Me—one of the first Hmong valedictorians in Northern California. Me—the first to graduate college in the entire Lee clan, the first to land a stable job, the first to plant a flag in the American Dream.

All the uncles and aunties clapped for me—hopes, immigrant prayers, and aching belief poured into my name. And that shit got into my bloodstream. Into my ego. Into my bones.

I was at the top of the world looking down at everyone else, thinking, *What's wrong with people? Success is easy. What's their excuse?* Look at me— I've made it. I'm brilliant. I'm unstoppable. I even wrote a whole damn book to prove it.

I really thought I was somebody.

Pride. The original sin. The king of them all. The one that always collects its debt.

And God—with His cosmic humor and perfect timing—said, *First of all, Son... that book was terrible. Second—sit tight. I'll show you what having it all really means.*

Flash forward ten years.

And I don't have shit.

Just darkness.

And always the earliest flight out of Reno.

I used to love this place. But you can't win with these little airports— they're like a ghosted ex-lover. Easy to get in, easy to get out, but there's always a catch. Always a layover. A lot of times, delays. I've been there when the lights shut off, the workers head home, and you're left stranded with your own miserable thoughts—no booze anywhere—locked in a silent war with a vending machine.

That morning, I woke up at 3 a.m., the witching hour for early travelers.

I got everyone ready—the newly crowned ex and my two boys—with passports, snacks, phones, chargers. Boxes and luggage were packed the night before in the back of the truck, sitting patiently like good little soldiers waiting for deployment.

I drove them to the airport. Parked in short-term. Three carts. Two boys. One dumb bitch. And one life quietly falling apart outside the terminal.

I held my sons long and hard, kissed them until I couldn't insist anymore. I told the big one to look after the little one, knowing he wouldn't. Told the little one to take care of his big brother, knowing he would.

"What's the most important thing in life, boys?" I asked—like I'd done every single day since they could speak.

They answered in that sleepy, rehearsed way:

"To be a good man. Don't lie, cheat, steal, or take advantage of others."

"And?" I asked.

A pause. Then:

"And don't be lazy."

They're still working on that one.

Good boys. Mama's boys. Still soft around the edges. Still shorter than me. Still innocent kids who thought the world was made of Minecraft and chicken fingers. Their voices hadn't cracked yet. In a few months, when I saw them again, the big one would shoot up taller than me, with a voice deeper than mine.

I hugged them one last time, breathed in the last trace of babyhood on the little one's cheek… and let them go.

Fist-bumped the bitch. Miss Bag-Fumbler. I would've given her the world. Gave her everything else. She never really had to work much. Maybe that was the problem. She always chose her own happiness over the family, and after years of struggling—for the safety of the kids—I finally had to let her go.

The kids chose her. I didn't want to drag them through the dirt. Didn't have any fight left in me anyway.

So she took them to Canada to live with her parents. At least I can visit anytime I want.

My in-laws—well, ex-in-laws—have apologized profusely since. I guess that's one of the reasons this whole book didn't turn into one big *Fuck You* throw-up.

"They're boys," I told myself. "It's good for them to struggle a little bit."

I hope I'm right.

As soon as I walked out of the terminal, the dam broke. Tears, snot, gasping, coughing—everything at once. I sat in the truck and cried the kind of cry that only happened one other time: when my little cousin Keng passed away from cancer over a decade ago.

The kind of cry that comes when the bottle and the soul are both completely drained.

I was a mess of grief on four wheels, trying to make it out of the stupid parking lot without dying.

I drank an entire bottle of Jameson that day. Didn't eat. Put the phone on Do Not Disturb. Woke up on the couch next to a sideways bottle and a broken mug—a crime scene of loneliness.

Yeah. Exactly like the country songs.

The sun came up the way it always does—soft and slow—spilling gold over a cold world that didn't want any of it. The cruel bastard just kept rising. The sky blushed with long colors through the blinds, but none of it reached me. I had already pulled the covers over my pounding head and stayed there for hours, until some annoying little bird started chirping out of sight in the backyard like it was trying to convince me the day would be gentle.

It wasn't.

Work started to slide. Complaints started to stack up. I could've cared less. My worst is most people's best. I was dead, and everyone was dead to me.

Mom and Dad were the only ones who noticed. They drove four hours from Willows just to drop off a pot of something home-cooked, or check if I was still breathing. Mom called every day. Texted every day.

Are you okay? Call me, Son. I want to hear your voice. I love you.

I'd type back, *I'm fine,* because anything longer felt like lying. It was exhausting to even do that.

Mom's voice kept me alive.

"One mountain a day, Son," she said—like it was a prayer.

Maybe it was.

Chapter 2: Rock Bottom

Life after that was cracking, peeling, blistering—like a miserable hangover that wouldn't go away, figuratively and literally.

Work became a slow funeral. Friends turned into ghosts with beer in their hands and pity in their eyes. Every kid I saw reminded me of mine—how they used to play, how they used to laugh.

Winters in Reno didn't help. Cold as ice. Gray as ashes. Long as regret. I used to be the life of the party—the guy who smiled through everything. Now smiling felt like bench-pressing the goddamn moon. Everything was heavy. Even being alive.

There were nights I woke up naked—couch, bed, floor, who knows—missing pieces of memory like a bad film reel. My friends cleaned up after me like I was their old drunk uncle.

Once, at a baby shower at a park, I took one shot of tequila and woke up naked, cut up, and home somehow.

Nope. It gets better.

Turns out I downed the entire bottle and tried to take a piss in the open at a park full of kids. My friends shielded me from a felony, threw me in a truck, and drove me straight home—straight into the weekly Friday night cul-de-sac block party.

The scene: me stumbling out somehow naked, disappearing into the bushes while my neighbors and their kids watched in horror. My friends dug me out like a wounded animal and threw me in bed.

The next morning, I asked the neighbors if it was that bad.

They all said yes.

One said, "Worse."

Time to get new neighbors.

The cul-de-sac crew had fallen apart anyway. The old man across the street used to snow-blow everyone's sidewalks. Then he got sick. He and his wife were already old, and things went downhill fast.

One Sunday morning, his wife rang my doorbell in a panic. He was in the hospital, and she had clogged the toilet. Water and shit were everywhere—bathroom, hallway, bedroom, all of it. I shut off the leak, sucked everything up with my shop vac, and set out fans. I checked on her every day until he came home.

They ended up moving out to live with other family and sold the house.

Not long after that, I finally patched up my own house, tossed all the junk, one memory at a time. Each thing in the dumpster felt like a lungful of fresh air. I kept as little as possible—sold the once-fullest dreams to someone else, gave her half afterward, and moved out of the new north.

Then I moved into a swanky 550-square-foot apartment on the old south side.

One bed. One couch. One large TV that never got turned on.

One lonely ghost.

No noise except my thoughts pacing the floorboards and the endless pouring of Jamesons.

Life became a dark gray loop—Groundhog Day, but without the comedy. Some nights, yeah, I thought about ending it. Not in a dramatic way. More in the quiet way you think about cliffs: staring too long, wondering how easy it would be to stop trying.

And then—on the first Friday after New Year's—the new owners of Kelly-Moore Paints held a mandatory company-wide conference call at noon.

They announced they weren't making enough money and effective immediately every single one of us—all 1,200 employees—was laid off. We had until the end of the month to figure out insurance, and that was that.

The company that bought us last year had been bleeding us dry and selling off assets anyway, so it wasn't a shock to anyone with a brain when the end finally came. The smart ones—like most of my old bosses—had already jumped ship long before. Most people—the ones who truly needed the job—were devastated. Some—the ones who were only there for the paycheck—shrugged, grabbed their few belongings, and took off without wasting another precious second or a wave.

And then there were a few of us who were just stuck.

People like me.

Strangely enough, I felt relieved.

Most of the managers and reps stayed until closing. We stuffed our belongings into bags and boxes, shook hands, and said the kind of sad goodbyes that marked the end of a long, strange chapter. Then we locked up the building one last time.

That night, I thought I'd be devastated—because I didn't know who I was anymore, what I wanted, or what the hell I was supposed to do next.

Not having a job made that clear as day.

I sat on the couch and tried to think like a responsible man.

Should I go home to my parents?

Start sending out résumés?

Make phone calls?

What did I even want from my life now?

Hours went by, and nothing came. Just a blank.

Outside the window: the same gray. The same cold. The same nothing.

And then the thought hit me—clean and clear:

Cali... you need to be somewhere completely opposite from here if you're ever going to think straight.

Opposite meant warm air.

Opposite meant sunlight.

Opposite meant white sand beaches in Thailand I never got to see.

I'd always wanted to return to Thailand, but every year there was some excuse—money, work, kids' braces, car trouble, family reunions. Year after year, life always found a way to get in the way.

But now?

Thirty-five years later?

No wife.

No kids.

No job.

No excuses left.

I thought of the quote from the movie *Friday*: "It's Friday. You ain't got no job. You ain't got shit to do!"

I chuckled.

Jesus.

So I took a deep breath, said fuck it, booked a one-way ticket to Thailand, and got on the first flight next morning. I didn't have a plan—not even a clue what I'd do when I got there.

But I knew exactly where to start.

Bangkok.

Chapter 3: One Way Ticket

I packed light. A couple pairs of shorts, a couple tank tops, a couple shirts, one pair of pants, three pairs of socks, and five pairs of underwear. I always go heavier on the underwear—you never know when you're one bad fart away from needing a fresh pair.

My first layover was in Seattle, so I headed straight to my favorite spot there: Salty's at SEA. Good as airport food gets—better than most. I went straight to the bar and ordered my usual: Salty's Famous Seafood Chowder Bowl, plus a double Jameson and a Michelob Ultra.

I always order double Jamesons—because singles don't do a damn thing, and at fifteen bucks a shot, why bother? A double is only twenty. Highway robbery disguised as a bargain, and I fall for it every time.

And of course, I get a Michelob Ultra, because that's what healthy people drink. A goddamn brilliant sales pitch.

I sat next to a very chatty blue-haired girl who asked where I was headed. Told her Thailand. She asked how long. Told her I had no idea. She thought that was pretty damn cool. She was on her way to yet another retreat somewhere in the Arizona desert. Definitely the type who "retreats" for a living. I told her I'd love to try one someday and asked for her number so I could check in on her adventures. Saved her as "Blue Hair Girl at Salty's." Never checked in. One more double Jameson for the road, please.

When we lined up to board, this big hairy dude with long curly hair past his shoulders and a black T-shirt walked up next to me, smiling like he knew a secret.

"Nice hat," he said.

I was wearing a cheap One Piece cap I'd snagged at the Asian store a couple months ago. Didn't know it then, but he was a massive anime fan.

I nonchalantly said, "Thanks, man," and we boarded.

I barely survived the flight to Tokyo. Twelve and a half hours of torture. I'd never been on a flight that long. Felt like everything behind me from my neck to my back down to my legs were alligatoring. Any movement felt like something was about to snap off.

I didn't have good headphones or earplugs. I'd already seen all the new movies on the entertainment system. I didn't have a book. No iPad. No neck pillow. No jacket. No backpack. Nothing. Completely unprepared.

The first four hours were uncomfortable.

The next four were brutal.

Thankfully, the plane was mostly empty, so I had all three seats to myself. At least I could throw my right leg across the seats every so often and find an angle that didn't feel like a hot screwdriver was drilling into it.

The big hairy dude sat a couple rows behind me. He immediately sprawled across his own three seats—hair everywhere, legs curled neatly, sleeping like a peaceful forest animal. Snored happily for most of the trip.

The goddamn bastard.

The last four hours felt like a death match. My ass was sore, my lower back was on fire, and my right leg felt like it was about to fall off. I paced the aisles, did squats and stretches like a lunatic, breathed like I was in labor, popped Advils, and ordered as many Jack & Cokes as they'd let me.

We finally landed in Tokyo at Haneda International Airport. I hit the restroom and ran into the hairy dude afterward. He said he was heading downtown to explore since we had a thirteen-hour layover.

I wished him luck.

My back was killing me. It was already close to 9 p.m., I hadn't slept a minute, and everything in my body was sore, burning, or breaking. All I could think about was a bed.

There was only one room left at the airport hotel. Expensive—over $250—but beggars can't be choosers.

The room was tiny. Tiny bed. Tiny bathroom. Tiny everything. Built for tiny people.

But it had a warm bidet, like they all do in this part of the world. The nicest bidets on earth are here in Tokyo. They've got a million buttons. Honestly, one of life's luxuries I can't live without.

I popped five Advils, set the alarm for 7 a.m., and crawled into bed. Spent most of the night trying to angle my right leg—now roaring with pain—into a position where it would stop screaming.

Eventually, I drifted off.

I woke up before the alarm, as usual. Haven't been a good sleeper in over a decade—five or six hours is about all I ever get. But my back felt calmer.

I took a shit, thoroughly explored my fancy bidet like a scientist, took a shower, got dressed, checked out, and headed back down to the terminal for the seven-hour flight to Bangkok.

Turns out I was seated right next to the same hairy dude.

That's when I finally got his name: Austin.

Austin told me he was spending a couple days in Bangkok, then heading to Koh Samui for a week, then Koh Phangan for the Full Moon Party. After that, his friend had the rest of their trip planned out. He had about $1,000 in cash for his whole month-long adventure.

"What's the Full Moon Party?" I asked.

"It's exactly what it sounds like," he said. "The biggest party in all of Asia. During a full moon. On an island. On the beach. I'm going there to meet my friend. She's already backpacking through Thailand with her brother."

Then he asked what my plan was.

I told him I didn't have one. The only thing I'd booked was three days in Bangkok—after that, I'd figure it out. I said the Full Moon Party sounded cool and asked if he'd mind if I tagged along.

"Nope."

Having someone to talk to made the hours fly by.

A beautiful bromance was born.

Chapter 4: Bangkok

We finally landed in Bangkok at Suvarnabhumi Airport at almost 5 p.m. I was so excited on the descent I couldn't bear to look down—afraid it would disappoint. Immigration was long but moved surprisingly fast—faster than anywhere else I've been.

Once we stepped outside, I breathed in a long-forgotten air. Hot. Humid. Sweet. Heavy in the best way. My sinuses opened up and my back softened instantly.

I spread my arms, tilted my head back, and exhaled, "I'm back. I'm finally back."

The air felt familiar, like it was talking right back to me. Welcome home. Like I had never left. My body knew it. Everything in me relaxed, and the back pain melted away on the spot.

Austin and I headed to the nearest ATM to grab cash. I took out 10,000 baht—which would be my usual draw, about three hundred bucks. He took out 3,000—his usual draw. He was on a budget. I wasn't. After a lot of pointing, asking random strangers, and Googling, we finally found our way downstairs to the Airport Rail Link.

Just as we were about to pay for our tokens, I saw shock wash over Austin's face. He started patting his pockets, digging through his backpack, spinning around in a panic.

He forgot his wallet at the ATM.

He sprinted back upstairs while I waited. Ten minutes later he returned and said he had to go to security because the wallet was gone. I sat down at a café and waited.

Another ten minutes later he came out holding the wallet, letting out the world's biggest sigh of relief.

Thai hospitality. Anywhere else and that wallet—and everything in it—would've been gone forever.

Hell of a way to start the trip.

His hotel was farther out, so we took the train to mine first so I could check in, drop my bags, and then head out to wherever he was staying to party. We got off near Sukhumvit—the heart of the city—and grabbed a taxi to my high-rise hotel, the Westin Grande.

The one thing I splurged on. Two hundred bucks a night for three nights, right in the middle of everything, just to get my bearings. Would've been a thousand plus per night anywhere in America.

Austin was impressed. Nineteenth floor, surrounded by skyscrapers, spotless grand room, and a fancy bidet that looked like it had its own engineering degree. The bidet and I will get to know each other well.

I put my stuff away and we dashed out—after stopping at 7-Eleven for our first Thai beer. A cold Chang. The first of many, many Changs.

We decided the fastest way to see the city was on the back of a motorcycle. So, we hailed two cheap moto-taxis. My first time ever on a motorcycle. I was scared shitless. I grabbed the driver with both arms like I was trying to cuddle him, my cheek pressed hard against his back. He peeled my arms off and made me hold the rail behind me instead.

So there I was—bare legs in shorts gripping the side of the bike above the hot exhaust, weaving and jolting through Bangkok traffic, heart in my throat. No helmet, like all cheap morons. Just hope, grit, and a locked jaw.

Meanwhile, Austin—one bike ahead—was having the time of his damn life. Legs spread wide, hands free, taking pictures on his phone, that long curly mane of his flapping wildly in the wind like he was born for this.

We finally got to his place—right in the middle of the famous, chaotic Khao San Road, where backpackers from every corner of the world collide. My legs were sore and burnt from gripping the hot bike so hard for damn near an hour in traffic.

He was staying in a hostel. Fifteen bucks a night. He had to stick his room key into a slot on the wall to get the electricity and aircon to kick on. Shared bathrooms, weird corners, slippery tile hallways, and random elevation changes like the whole place was built on top of a stack of Legos.

He dropped off his stuff. By then it was dark, around 8 p.m., but you couldn't tell when we stepped outside. The entire street was glowing—neon signs, bars stacked on bars, music blasting in every direction.

Austin had a grin stretched from ear to ear, walking slowly, head on a swivel, taking everything in.

Me—I was in my own world.

I finally made it.

Jesus Christ—I actually made it.

Khao San Road. I didn't know what it was going to be, but it was everything I dreamt it'd be. For years it lived in my head like a faded rumor, like some postcard someone else got to open. A place for the young, the wild, and the free. A place I'd given up on ever seeing.

It was an overload of the senses that felt like a descent into something unreal.

The air hit me first—hot, humid, sweet with mango and charcoal, fried chicken, and incense—all tangled together like a disco-ball of smells detonated in my sinuses.

And the lights…They didn't shine. They shimmered. They danced and winked and glimmered.

They reflected off puddles from a rain that ended hours ago but felt like it's still falling in some corner of my mind—only to be scattered again by an endless jungle of neon glitter.

Music blasted from every direction—EDM to the left, reggae to the right, and some dude with a guitar screaming his heart out straight ahead.

And the crowd… "busy" isn't the right word. It was shoulder-to-shoulder, body-to-body madness. A swirling river of faces. Backpackers with sunburned shoulders. Thai students laughing like the night had no expiration date. Hippies and hipsters—young and old—who never made it home. Vendors smiling like they've seen every kind of human on earth come stumbling through.

Girls in bikinis grabbing your arm, groping your balls, slapping your ass, tugging you toward their bars—you practically had to beat them off with a stick. It was fantastic! Guys twirling signs with erotic arrows.

A kid tried to sell me bracelets. "Sorry, no."

A dude shoving a scorpion-on-a-stick in my face: "Just one."

Tasted like burnt chicken. Crunchy, salty—burnt or not burnt enough.

Rushed past the bugs because I knew they tasted the same as burnt chicken. Wished for them when I was little, didn't have them when I was sick of dried sardines. Now, not a chance.

I'm asked if I want a tattoo.

"Memory forever! Tyson tattoo?"

God no—maybe—who knows anymore.

For a moment, time stopped. I wasn't tired. I wasn't old. I wasn't broken. I was… alive.

I felt like the seven-year-old kid from the refugee camps in the mountains, the kid who once dreamed of seeing Bangkok, and now here he was—walking right through the heart of it.

The universe—slow bastard that it is—finally answered.

All the heartbreaks, all the grinding, surviving, saving… they led here. To this street that smelled like the beginning of something hopeful wrapped in fish sauce.

I walked slowly, afraid to blink, afraid to face.

I heard languages I couldn't place. Heard a laughter that felt eerily like mine—like something I lost long ago and finally found again.

Every step felt like redemption.

Every flash of neon, a reminder: *"Look what you survived long enough to see."*

I grabbed Austin's shoulder. He turned.

I started: "And away—"

He finished: "—we go!"

And just like that, the adventure officially began.

We took a shot of Maker's Mark and chased it with a Chang at the first bar. I prefer Jameson when I'm alone, but when I'm with people, I make the bartender's life easy.

We found a louder, wilder spot right after. I ordered each of us a Maker's bucket—a full-on gallon bucket with ice on the bottom, a medium bottle of Maker's, a Coke, a Red Bull, and a couple of straws.

We dumped in everything, swirled it like a witch's brew, and started chugging.

Khao San Road cracked open its jaws and swallowed us whole.

Chapter 5: Round Two

The next thing I know, I'm waking up in Austin's room. In his bed. Fully clothed. He's passed out next to me, snoring like a goddamn dying lawnmower. I found my phone on the counter and checked the time.

10:30 p.m.

The same damn night.

I don't remember a single thing after we started chugging those buckets. Not a flash, not a sound, not a hint of how the hell we got back. Just black. One minute I'm swirling Maker's, the next I'm in some dude's hostel bed like a confused house cat.

I sat up and did the standard post-blackout survival checks: Phone? Wallet? Asshole? All intact. Nothing sore.

I shook Austin awake.

"Bro… what the fuck happened? Do you remember anything?"

He rubbed his eyes, groaned, and said, "Nope."

Perfect. Two grown ass men blacked out before 11 p.m. in the most chaotic place on earth. Outside, the music was muffled and getting even louder.

"Well," I said, "the night's still young. Round two?"

"Let's go."

We agreed we needed a $7 one-hour Thai massage first. No happy endings. Just repairs.

Then we hopped on a tuk-tuk to Chinatown to grab street food—hands down the best and busiest Chinatown I've ever seen in my life. Street food Mecca of the entire goddamn world.

It was chaotic in all the right ways. Hectic. Loud. Glorious. Every smell punched you in the face and then tried to kiss you after.

We tore through endless skewers—pork, chicken, octopus, and grilled things that I'm sure weren't FDA-approved, but tasted like they were blessed by the angels themselves. I don't know why, but I wanted to sing a song.

Some were so good I proposed to one of the skewer ladies. That's according to Austin anyway. I don't remember bending the knee, but I wouldn't put it past me.

There will never be a Chinatown better than this. And if there is, I swear to God—I'm coming for you, and I'm going to eat you.

Then we hit Khao San Road all over again—dancing with locals, travelers, randos from everywhere. Most bars had nobody on the dance floor. Just standing around.

Then I stepped in. Austin joined. Then the girls. Then the guys.

And suddenly I'm buying shots for everyone like I was Oprah Freaking Winfrey. We skipped the buckets this time.

Didn't matter. The world still went black.

We woke up around 10 a.m. in his room with the meanest hangover imaginable. It takes a lot for me to get a hangover, so I guess we did it right the night before.

Austin took a shower while I waited. When I checked my wallet, it was completely empty.

Figures.

Time to pull out another 10,000 baht and keep moving.

We grabbed a tuk-tuk to my hotel so I could shower and explore the amenities a bit before tackling the one mission I had to complete: eating the world's best noods at Boat Noodle Alley, about ten stops north on the MRT Sukhumvit line.

We finally made it to the famous Boat Noodle Alley—a dream of mine, a true nood-lover's bucket-list item.

In the old days, vendors would glide along this canal in wooden canoes, serving tiny bowls of noodles straight off the boats to the locals—hence the name. My cousins had all been here years ago, posting pictures of ten-bowl stacks like some sort of Hmong noodle Everest.

But we were so brutally hungover we could barely see straight. All the poison from last night was still leaking out of our pores, threatening to hurl itself onto the side of the canal with each bite.

We mustered enough strength to slurp down just two tiny bowls each before tapping out like the little sweaty bitches that we were. Tiny crumbled napkins everywhere on the table.

I came, I saw, and I conquered... two measly bowls.

A sad showing for such a sacred place. But hey—mission accomplished.

We swam at my rooftop pool and took in the views for the rest of the afternoon. Then we changed into dry clothes and went out to explore downtown Bangkok—which, in our case, meant barhopping through the red-light district of Nana Plaza, where the ladyboys roam.

Austin wanted absolutely no part of it, but I insisted. I had to see it for myself.

I dragged him into this three-story mall of ladyboys with buzzing neon bars stacked on top of each other. We found a ginormous spot where at least twenty-five girls were dancing on stage.

Some were clearly boys—the bulge or the Adam's Apple gave it away. But some?

Impossible to tell.

Delicate features, perfect proportions, pretty—even stunning.

The waitress brought us some Changs and a couple shots of Maker's and asked which dancer we wanted to pick. I pointed at one even Austin had to admit he couldn't identify. He was visibly against this entire adventure, but I didn't care.

The waitress waved the dancer down to sit with us. She ordered a watered-down Red Bull Vodka.

I asked if she was really a ladyboy.

In broken, husky English, she said yes.

She proudly pulled her panties down just enough to reveal the beginning of a penis about the same size as mine. It was dark, so I couldn't tell if hers was cute like mine or not. I asked how she hid it so perfectly on stage.

The secret: two pairs of underwear. The first has a built-in cameltoe. The second holds everything in place.

Made sense.

The Hangover movie curiosity: satiated.

Austin looked like he wanted to evaporate. He was trying hard not to watch any of my shenanigans. As soon as we finished our drinks, he bolted for the exit—damn near sprinting.

Right as we were leaving, an angry Russian guy was screaming at security as they tossed him out. Behind them stood a ladyboy. The Russian just found out the "hard" way and was not happy about it.

I laughed hysterically.

Even Austin had to admit it was hilarious.

We hit a few more bars on that street, grabbed some drinks, and danced all night with strangers from every corner of the world.

We hit a rooftop bar. I showed up in a tank top and got denied, so I marched across the street and bought a breezy, colorful island shirt from the ladyboy running the shop.

Light, bright, and my size—a true small. Proper smalls made by people who know small. In America it's always a gamble: mediums pretending to be smalls, "schmediums" with an "S" on the tag.

Four bucks each only. I didn't need to negotiate.

She had twenty of them on the rack. I told her to save them all for me. Came back the next day and bought the entire stack.

Eventually Austin had to take off—he had an early morning flight to Koh Samui.

I found a tiny outdoor bar with a couple of tables and chairs right next to my hotel to relax. They were playing **Sabai Sabai** by Bird Thongchai from the old school days—songs my dad and my uncles would play all the time. Seemingly the national anthem of the Thai people, which means *Relax Relax*.

I sat there drinking Changs all night with a young, beautiful waitress who laughed at everything I said, even the things I didn't mean to be funny. There was a slight breeze. It was the perfect Bangkok night.

Jesus Christ.

The next day was dedicated to serious noods. A vengeful pilgrimage. Retaliation against the nood demon kings that humbled us the day before.

Noods, noods, and more noods.

I explored every alley and corner within walking distance. Threw in a Chang or ten at a few bars in between to cool down from the heat and the spice.

But that night, something shifted.

I met a British bloke named Patrick, visiting from Vietnam. Same story as so many wanderers out here: girlfriend back home dumped him, he took it hard, came to Thailand to "find himself or something," and—unlike most—actually did.

He loved it so much he stayed. Became an English tutor, then a tour guide on one of the islands after a friend hooked him up.

Five years later, that same friend moved to Vietnam to start another tour guiding business—Thailand was getting expensive anyway—so Patrick followed. Been there five years now.

He was finally back in Thailand to see old friends from home who were visiting—his first return after a decade of bouncing between islands, borders, visas, and a life lived out of backpacks.

I listened to him talk and thought, *I could do that.*

Hell, I could do it even better.

The longer I stayed, the tanner I got—practically a local already. A couple days here and I was already picking up the language I'd lost. The food—still the best in the world.

And I can pull off any goddamn job here to support myself.

For the first time in a long time, I saw a path. A small one—but better than the nothing I'd had for years.

Optimism.

What a strange feeling…

Chapter 6: Koh Samui

I should've bought a ticket earlier, but I got too serious about my noods during the day and didn't stumble back to my hotel until three or four in the morning. When I finally woke up, I didn't have many options left, so I played it by ear—grabbed a taxi to the big airport and tried for a last-minute standby.

Nope. Not a chance. Not in the high season. Another lesson learned.

I ended up waiting at the airport—six long hours burned—before I finally scored a flight to Koh Samui. Thank God, too, because it was one of the last flights they had. I texted Austin the play-by-play the whole time.

At last, I got on the goddamn plane to paradise.

The takeoff was long and delayed on the runway, like all of them here. Especially at this airport. Loud. Hot. Sweaty.

A perfect prelude to the upcoming party.

But the descent was one of a kind.

It's one of the biggest islands in Thailand. It looked like the most island thing I'd ever seen to date. It reminded me of Maui in the States—sitting a stone's throw off the peninsula near Surat Thani Province, a gateway to a whole sea of even more magical places.

It came into view slowly, shy at first, like it wasn't sure whether to reveal itself to me just yet. Not too much sun. A little mist. Clouds for mystery. Then the beauty started stacking—palm-lined beaches blending into long green humps of jungle, tin rooftops shimmering beside temples painted in gold leaf, every building and structure—every color—stitched perfectly into the coastline like an omnipotent being placed them there by hand.

From my window seat, it felt like the whole place was rising up to meet me. Even the clouds seemed to part out of respect. I pulled my phone out and took endless pictures like a man trying to prove a miracle happened.

The closer we dropped, the more impossible it became. The ocean shifted from deep sapphire to electric turquoise that didn't look real. It looked Photoshopped. Cooked up in some tourism brochure to lure suckers like me into paying too much for a hotel with bad plumbing. But this wasn't a brochure. This was coming straight at me at five hundred miles per hour.

Behind the coast, the jungle rose in a perfect, unstoppable wall of jade—dense, ancient, full of secrets. A green so alive I swore it was breathing. The kind of green that makes you forget every beige office cubicle you ever sat in, and every fluorescent light that ever devoured the color right out of your life.

The place didn't simply appear. It unfolded. Like an emerald flower.

Boats—big ones, little ones, fishing boats—drifted lazily like they had nowhere urgent to be, like they were mocking me for thinking the world was supposed to be lived any other way.

Farther in, I could see tiny roads curling into the hills where houses clung to cliffs. Patchwork roofs. Laundry lines flapping. Smoke rising from outdoor kitchens. Right next to sprawling, majestic, English-speaking, uppity hotel chains.

Messy. Imperfect. Human. My face was glued to the window.

The plane banked a little left and the whole cabin gasped in unison like it was rehearsed. The wing dipped toward the sea for a second, giving everyone a crisp panoramic shot of the island's northern tip—a long slice into the ocean like a Swiss Coffee knife, dotted with beach shacks, bars, and tiny plastic chairs where entire fortunes of happiness were traded for the price of a beer.

That's when I felt something true swell up inside me. Something small as my thumb. Something quiet as a low hum. Something I'd been too numb to feel for years.

A firefly… of hope.

Or something that felt suspiciously like it.

Whatever it was, it had been a long time since I wasn't thinking about the past—not my failures, not my heartbreaks, not my mistakes, not the colossal, collapsing life back home. I wasn't thinking about what I'd left behind, what I'd lost, or what I couldn't fix.

I stared at this green paradise—rolling hills lifting themselves out of the white-sand edge of the sea—and thought that maybe, just maybe, something in me might rise with it.

This was the total opposite of the cold, dark, miserable place I'd been running from.

I knew that somewhere down there, I'd find the slightest bit of… direction.

Something.

Anything.

Then the landing gear dropped with a thunk. The engine changed pitch. The runway extended out into the sea like a dare. The whole plane tilted forward, and the island grew larger, sharper, louder in my window.

And for one clean, perfect moment—right before touchdown—when the whole world hung suspended between sky and land, something whispered in the back of my skull:

You'll never be the same again, Cali.

Then the wheels hit the ground. Hard. Jarring. Real.

The engine roared in reverse, the wings trembled, and the cabin silently cheered like we'd just survived something heroic.

All I could do was smile.

And as the plane rolled to a stop, still humming from the descent, I knew one thing for certain:

I already loved this place.

We got out and shuffled onto the shuttle bus headed for the terminal, the way all island airports seem to do it. I went through their little doors, took a piss, grabbed the first taxi I saw, and took pictures the whole way down to Lamai Beach—half awe, half familiarity.

I called the bastard.

He answered.

"Let's fucking go! Right?"

"Right!"

I went straight to Austin's hotel to grab a Chang and check out his shack right on the beach. You had to stick the key card into a slot on the wall to get the electricity to turn on—only one key, of course. The first of many places like that. Tiny low bed, almost on the floor. Tiny bathroom with a bum squirt gun, but at least it wasn't shared this time.

And honestly, for ten bucks a night, you couldn't complain. The place sat right on the beach. You could wake up, open the door, walk thirty feet, and dive straight into the ocean.

Tell me where else in the world you can do that for ten dollars.

Then we headed over to check into my much nicer, unplanned hotel—a couple blocks east. Cheapest I could find on Booking.com.

They greeted us with butterfly pea flower tea—beautiful, purple, and delicious.

After that, if a hotel didn't have the purple tea waiting for me, I'd be unhappy.

Austin grabbed one, loved it, and immediately asked for seconds even though he wasn't staying there.

They didn't mind.

Let me tell you—it was the best $35 hotel I've ever stayed in. Beats most $500 hotels in the world, bar none. Fancy curtains on the bed and a fancier bidet. Oh yeah. They even gave me two keys so I could keep one in the slot and let the aircon run. Every time I stepped back in from that 100% humidity, the room was already an icebox—my preferred way to nap or sleep.

For short trips? That's luxury, baby.

And it was right next to the goddamn ocean.

Stop it, man.

That night we grabbed a tuk-tuk and headed to the party strip near downtown to catch a Muay Thai fight—first thing on the agenda. Austin had always wanted to see one in person.

Me too.

Back in the camps, my dad and all my uncles would crouch around a small, dirty radio and listen to Thai announcers call the fights like it was war poetry.

Spinning elbows! Flying knees! Roundhouse kicks!

So yeah—I was *beyond* excited.

We got there early, scoped the place, bought our tickets, then wandered down the street looking for a bar.

Of course, Austin sniffed out a reggae bar instantly. He's a huge reggae guy—loves Sticky Fingers.

Now I do too.

We knocked back a couple jungle-named drinks, danced without a single care in the world, and vibed to the music before heading back for the main event.

The fight was everything I'd imagined it would be.

Loud, enthusiastic announcers. Traditional Thai music screeching in the background. Ritual pre-fight warm-ups like a sacred performance. Thick sweat hanging in the air. Crowds cheering. Legal *and* illegal gambling happening all at once.

Smoking-hot ring girls. Hard concrete benches. The whole place rough and unforgiving—like those elbows and knees.

Young white dudes knocking out old Thai guys. Brutal.

But the women?

The women had the best fights by far.

Mission accomplished.

We bar-hopped a bit on that busy street, dodged girls left and right, then grabbed another tuk-tuk back past three 7-Elevens and a left toward the Lamai bars, much closer to our hotels.

That's how directions work on the island, I learned—count the 7-Elevens and sprinkle in a few lefts and rights until you get there.

We ended up at our favorite reggae bar, run by a tiny Thai lady and her white husband who only emerged from the back occasionally. Played pool. Knocked back Changs and cheap whiskey—our specialty.

Then we wrapped the night with a midnight swim in the ocean.

Our nightly ritual.

Another great night sealed.

I woke up early, as always—always waiting on Austin to finally stir. By then, I'd already slipped into what would become my daily ritual:

Wake before the sun.

Lay in my beachside hammock and watch it rise.

Knock out 100 dips, 100 pushups, 100 squats—right there on the sand.

Float in the ocean so salty and buoyant I barely had to move. Sometimes I'd doze off in the water and only wake when the shore nudged me or a wave slapped salt up my nose.

Then I'd wander down past Austin's hotel to the little local breakfast joint and eat a simple bowl of minced pork porridge—ninety cents of perfection.

After that, I'd head to my favorite old ladies for a $7 one-hour Thai massage to reset whatever the night had broken.

And then I'd park myself at Austin's bar, wait for his slow ass to wake up, and order my usual trifecta:

— a Bloody Mary for some hair of the dog

— a coffee to wake up the ghosts

— and a watermelon smoothie for electrolytes

All for five bucks.

Austin finally got up, and we spent most of the day exploring every day-market we could find—downtown and all around.

Tuk-tuk rides. Changs. Red Bulls. Street food. Beaches. Bars.

Rinse, repeat.

We ended up back at the reggae bar from the night before. Played pool for hours. Let the reggae drift through us. Just… breathed. No rush. No plans. No care.

Jah was smiling.

We closed the night the best way we knew how: a quiet dip in the ocean under a warm island sky, a smile on our faces.

A perfectly slow, perfectly unrushed island day.

I wouldn't have it any other way.

On the third day, we walked across the street and rented a cheap motorbike from a wonderful lady named Mook. She sold drinks too. I ordered a fresh coconut her husband had literally just picked from the backyard—huge coconut trees poking out behind the house like towering guardians. Grabbed two Changs for the road and asked if she had any whiskey.

That's when we got introduced to Mekhong—the cheap Thai rum that tastes like whiskey if you squint your taste buds just right.

We took a shot each, liked it, and instantly added it to our little rotation so the Changs don't get lonely.

The motorbike was only 250 baht for the day—seven bucks—and already filled with gas. Helmets on this time. Austin hopped on like a pro, turned the bike around smooth as butter—he's been riding his whole life.

I hopped on behind him and clutched his shoulders like his Asian boytoy.

We took the winding road north through town to the top of the island to explore Fisherman's Village. Walked the beaches. People-watched. Grabbed smoothies. Sipped a couple Changs. Checked out art shops. Ate BBQ skewers at the day market. Browsed wooden carvings shaped like penises for reasons unknown to many. Wandered through narrow alleys.

Then we looped back through town, found another night market that hadn't fully opened yet, and headed back to our beach to cool off and wait for it to pop off, as Austin said, so we could return when the magic started.

We grabbed a couple Changs and sat at a table. To our right was a young, short, cute blonde with braids, eating something with her larger friend—maybe her sister. The friend was bundled up from head to toe with a hoodie over her head, face on the table. She had to be cooking under all that fabric in this heat.

Blondie glanced over and I waved and smiled like a goof. She gave me a half-smile, then turned back to her not-at-all-amused friend.

Then I spotted this young white guy in the corner—colorful island shirt, XXL minimum. Built like a Greek god, arms like tree trunks. The dude was a tower. Even sitting, his upper body was damn near my standing height at the table. Wide straw hat. Face shaded like a final boss.

You didn't need details.

The dude radiated lady-killer from thirty feet away.

We finished our beers, hopped on Austin's bike, and headed back to the night market. It was already alive and massive—sprawling and loud—and it had only just opened. It rivaled even Chinatown. Every kind of seafood, meat, veggie, soup, pastry—cooked every possible way: grilled, fried, boiled, steamed, skewered, torched.

Overwhelming in the best way.

Austin nibbled on skewers and random small plates. I, meanwhile, locked onto a massive cauldron with a line wrapped around the stall. I recognized the dish immediately from my cousins' old Thailand trips—a towering pork bone stacked high with a mountain of volcanic herbs, chilies, and spices spilling down the sides like an eruption.

My mouth watered. I caved. Ordered it. Went full caveman.

There's nothing neat about eating it—no clean way, no polite way. Just spice and sweat, fingers gripping hot bone, teeth ripping and pulling, lips burning, tongue slurping broth that tastes like hell and heaven at war. I demolished half the napkin container trying to keep up.

By the time I was done, all that remained was a stripped-clean bone and a small mountain of shredded tissues.

We returned victorious—and repeated our nightly ritual.

The next morning, Austin said he'd overheard someone at the bar talking about a waterfall.

"Sweet," I said. "Let's check it out."

"Yeah," he replied, "but I don't think we can make it with you behind me. It might get bumpy. You might fall off."

I stared at him, disappointed.

"Soooo… how exactly are we getting there then?"

He shrugged. "You need to get your own scooter, man. Besides, you look like a little bitch hanging onto me. I don't want people thinking I'm carrying around my gay Asian boyfriend all over the island."

"What the fuck, bro!" I shouted. "I've never been on a fuckin' bike before and you want my first ride to be some bumpy-ass road to a waterfall with you? That's messed up!"

He pointed at a blonde girl in a skirt riding by on her scooter. "Don't be a bitch. She's not a bitch."

"You motherfucker…"

But he wasn't wrong. And I was tired of clinging to his back like a nervous, sunburnt koala anyway.

So, reluctantly, we walked across the street to Mook's. Grabbed smoothies and told her I needed to rent a bike too. She asked if I'd ever ridden one.

"No."

She smiled. "No plablem. Vedy easy. Plactice ova thea."

She pointed to the empty dirt lot beside the shop.

I was shaking like a chihuahua in winter.

Austin taught me how to ride.

"It's like riding a bike, bro," he said. "Easy on the gas. Let the bike do the work."

I took a deep breath.

If that blonde could do it, so could I.

And just like that, my little bitch era officially ended.

It was time to be wild and free—exactly what this whole mission was supposed to be about, right, Cali?

Riding my own scooter felt like the first necessary rite of passage toward real freedom.

Finally comfortable, we took off toward the waterfall. It was scary at first but thank God the road was nearly empty. I had room to breathe.

And he was right—it really was just riding a bicycle, with a louder engine and deadlier consequences.

The only time I fell was when I tried to park on the small rocky slope near the waterfall—because I followed Austin exactly.

He turned around and yelled, "Bro, I told you to follow me, but not do everything I do! Do what feels safe!"

Fair enough.

Just a tiny scratch on the bike, thank God. I'll tip Mook extra. It was only my ego and my left side that were bruised.

I can walk it off.

We marched down toward the waterfall through jungle vines and rocky cliff paths. It was tougher than it looked, but worth every damn step. Only one couple was there, and they took off the moment we arrived, leaving the whole place to us.

I jumped straight into one of the pools.

The water was indescribable—fresh, cold mountain water… on a tropical island.

I tilted my head back and drank from the firehose relentlessly pouring over the rocks above.

Delicious. Refreshing. Healing.

Austin jumped in too. We sat there until our fingers turned to raisins, grinning at each other. We kept repeating, "There's nothing better than this."

And there really wasn't.

Eventually, we climbed back up to the parking area, which overlooked the entire northern end of the island and the sea beyond it. We took panoramic photos from every angle like tourists possessed.

"Stunning" doesn't begin to cover it.

Austin was in awe.

I was… home.

I was born in mountain jungles like this. High up. Remote. Wild. My body recognized it instantly. It felt like something deep in the earth exhaled and whispered in a voice like my grandpa's:

Welcome home, son.

There was an old man relaxing in the shade selling water, sodas, and incense, so I bought a couple waters and some incense from him. We walked over to the biggest shrine on site. I showed Austin how to pray: light a couple sticks, kneel, put the incense to your forehead, thank God, thank your ancestors, ask for blessings and protection—for my family and kids back home, for myself, for my best friend Austin traveling beside me—then bow three times and place the incense in the rice bowl with all the other burnt ones.

And after that… I didn't say much. I didn't need to.

I said goodbye to Grandpa, and we continued on.

Chapter 7: The Three Amigos

On the way down, we stopped at a roadside bar and grabbed a Chang from a beautiful tatted-up bartender. I asked her to marry me so I could stay here with her and help her with the dishes. She ate it up. I promised I'd come back and see her again, then we cruised back down onto the main road and rode all over the island. I even let one hand off the handlebars to record myself ripping through the jungle like some badass who absolutely did not almost die a couple hours earlier.

Later that afternoon, we returned to our usual bar, grabbed two Changs, and sat on the beach. To our right, about ten feet away, I spotted the cute, braided blonde from yesterday—this time sitting alone. She glanced over and gave a small half-smile. I returned the same.

Curious, I walked over.

"Hey, what happened to your friend?"

In a slight German accent, she said, "She's sick. She's been sleeping in the room for two days."

"That sucks," I said.

She nodded.

I asked what they were doing on the island.

"We came for the Full Moon Party."

"No way—so are we!" Like I didn't know half the island was here for the same reason.

I introduced myself. She told me she was from Zurich. But I decided she looked more like a Berlin.

I called Austin over, and the three of us talked. She said she was leaving in two days—the day before the party. I told her we were leaving tomorrow to beat the chaos. She said we were lucky.

I asked if she wanted to grab dinner and come party with us instead of babysitting Ol' Sicko in the room. She lit up. "Yes! I'm not going to ruin my vacation because of her!"

"Cool. Give me your number. I'll text you in a couple hours."

We high-fived her, took a dip in the ocean to cool off, went back to the bar, and ordered another Chang.

Then the tall blonde dude—the Greek god from yesterday—walked over and asked if he could sit. Young Arnold with a straw hat.

"I saw you talking to her," he said. "She's German. I overheard them yesterday."

"Yeah, she is. Sit down, bro."

And that's how we met Daryl.

Swiss. Obviously.

"You should come to Switzerland, Kao!" he said. "Guys like me are everywhere. You'd be a unicorn. All the girls would love you—especially your personality!"

Sure, dude. Sure.

I didn't know if I was supposed to feel flattered or deeply concerned.

He was going to the Full Moon Party too. Same ferry tomorrow. Same plan: get there early before the cataclysm of backpackers, hippies, hipsters, and partiers from all over the world.

We asked if he wanted to roll with us. He lit up like a kid on Christmas. We exchanged numbers.

A group chat was born: The Three Amigos.

Night fell.

We all met in front of Austin's hostel and headed out. They all stayed at the same damn place, so it was convenient. We hit the night market. Everyone grabbed food from different stalls.

I found a nood stand—and heard the owners speaking Hmong. Hmong people owning noodle shops? Normal. But my people all the way down here—on an island? Wild. It almost brought a tear to my eye seeing how far we've made it, and how we're still out here trying to thrive in every corner of the world.

I ate. Delicious, of course. Duh.

There was a live band next door. We checked it out. I ordered drinks for the whole crew.

At some point, Berlin and Austin swapped tops—Austin in her tiny string top, Berlin drowning in his oversized red T-shirt like a grandma. Then me and Daryl swapped. His XXL shirt swallowed me whole, and my tank top stretched over his big ass body like it was fighting for its life. Then we swapped again, and I ended up wearing Berlin's string top the rest of the night.

Chaos ensued.

Berlin and Daryl got up on stage to sing "99 Luftballons." She carried the whole thing. Daryl just hopped around like a golden retriever on crack. I watched Berlin soften toward him right there.

Later, Berlin started whipping me with my own belt. Then this stunning brunette in a floral dress—Wales—sat next to us. I asked if she could do better. She tried. The belt nearly snapped. Didn't feel a thing.

She had a gay friend with her. I pretended to care about his name. I didn't. I asked if he was her boyfriend just to make sure before I pulled her away to dance.

More drinks.

Then body shots.

The girls took turns taking shots off Austin. Daryl did too. I refused. I waited for my turn, but it never came. Fuckers.

We got so loud the people in the crowd—who were cheering us on at first—started glaring at us and shushing us. Then the band got visibly annoyed because we had the back rows riled up like a riot. Then security kicked us out.

We took the blame so Wales and the others wouldn't get booted too. I paid the tab blind. Didn't care.

One of the greatest days and nights of my life.

I got Wales's number. Kissed her like a drunk. Didn't remember if it landed or not. Told her I'd call. Never did.

However, I have a type now. Tall, slender, sophisticated, fun. Jesus Christ.

We went to another bar, changed back into normal clothes, and ordered more drinks. Austin stepped outside for a cigarette and vanished. Wouldn't answer his phone.

We retraced our steps along the beach back to the hotel. Daryl suddenly ripped his clothes off and yelled, "Skinny dipping!" and sprinted into the waves. Berlin followed. I stayed dry like a fully grown adult.

Eventually we found Austin passed out on the sand in front of his room—waves splashing up on his feet. Slurring. Pissed. Missing his phone.

We searched everywhere. Almost gave up. Tried to get him back to his room. He refused. Just started smoking cigarettes. They'd had enough of him at that point.

But I don't leave my brother behind for anything.

Then I had a thought: Maybe he rolled over it, the fat bastard.

I dug into the soft sand like a dog at the exact spot he'd been passed out in.

A foot and a half deep later— I found his damn phone.

He was ecstatic.

I finally tucked his ass into bed.

Maybe I was his good luck charm. The man almost lost both his wallet and his phone in one week.

Berlin made an excuse not to sleep with her sick friend. Daryl invited her into his room. I waved them goodnight, walked back to my hotel, took a midnight swim, and drifted off… with a smile.

Chapter 8: Koh Phangan

The next morning—thanks to a few last-minute changes in our grand, almost nonexistent plan—we ended up catching a snorkeling trip to Koh Phangan instead of the usual ferry.

Daryl came bouncing over like an excited puppy, saying he'd just talked to some guy across the street who was selling "cheap snorkeling tours" that stopped on the way to the island.

"Only 1,000 baht!" he yelled.

The cheapest ferry alone was almost 500.

"So we get a tour and a ride?"

We looked at each other.

"Fuck yeah!"

We sprinted across the street and somehow booked the last three spots.

In hindsight, that boat was already packed to the gills—shoulder-to-shoulder, almost-spilling-out-the-sides packed. The guy must've taken one look at us and thought: Eh, squeeze in a few more dumb tourists. Why not.

And away we went—crammed in the very back, sunburned, laughing, and absolutely convinced we'd just beaten the system.

A couple island stops. A couple stupidly beautiful beaches. Then the standard buffet-style Thai lunch on a majestic thing of an island—Koh Tao—at some half-shaded picnic spot where a group of young European girls sat a few tables over, absolutely infatuated with Daryl from afar.

You could see it in their eyes. Every time he stood up, stretched, or even breathed, they whispered and giggled like they were watching a goddamn movie come to life. Austin and I shook our heads. Daryl smirked, pretending he didn't notice.

The bastard noticed.

Immediately he went to talk to them after lunch, and it was like watching fangirls meet their idol in real life.

The third and fourth snorkeling spots got canceled because the ocean turned angry.

On the way to Koh Phangan, the waves got even rougher. We were jumping waves at one point. A kid threw up his lunch, and I followed right after him—feeding the fish.

The boat guys were cool about it. Jumped into action, poured cold water over our heads, told us to look straight ahead and breathe.

Pros.

We finally got dropped off at the main pier and negotiated a tuk-tuk to Daryl's hotel, right in the middle of Haad Rin Beach—the epicenter of the Full Moon explosion coming in a few days.

I managed to find a nice room with a bum gun right next to the ocean at the same place for $55, but the catch was brutal: the island had been booked solid for months by real travelers who actually planned things.

Especially for an event like this—if you don't plan, you're pretty much shit out of luck.

The nights before, during, and after Full Moon were completely sold out. I'd have to find somewhere else after that night. The only place with a bed left anywhere close to the boys was a hilltop resort a couple blocks out of town—$300 a night.

Ridiculous. Especially here.

But beggars can't be choosers. At least I'd have a room and a bum gun, and it was a walk I could make.

Then we headed straight into the center of all the action so Austin could check into his little hostel.

We grabbed a quick bite at a cheap, simple restaurant. By this point, the trio already knew each other's orders by heart: Daryl got some kind of curry. Austin went for something fried with rice. And I was obviously getting noods.

We partied that night the way we always did—three dorks dancing, drinking, singing, laughing our asses off through the chaotic Haad Rin alleys.

At one point, somebody had the brilliant idea to take an 800 baht tuk-tuk ride up north to find some mind-altering drugs at this hippie party we might need for the Big Party. Austin said he knew a guy who knew a guy who might actually be there.

So he called the guy, who called the guy, and off we went.

It started raining on and off during the ride.

We eventually found the party, but the guy didn't answer his phone for a while, so we danced in the rain first—then squeezed into some loud, sweaty bar packed with drunk, sticky partiers bouncing up and down in unison to some annoying electronic thumping.

Hippies with dreads were selling trinkets and random handmade shit everywhere.

I started scanning the scene for the most fucked-up dudes to ask if they had anything.

Usually works.

Nope.

Everyone was just stoned out of their minds.

Finally, the guy called Austin back.

We met up with him, but he didn't have shit—just a huge French spiel about how handsome he was, how handsome Daryl was, how Austin and I have no chance in life, how he constantly must fend off marrying beautiful ladies who want to fly him out to every corner of the world…

Then back to how handsome he was again.

I thought: I've finally won at life. I've met a handsome French drug-dealing magician.

What a goddamn waste of time.

Meanwhile, Daryl had gotten stuck talking to a hippie girl, as usual. I asked if he wanted to stay with her while we went ahead.

We started leaving.

He ran back a minute later, ticked off at me, saying I'd cock-blocked him because she walked off right after I said that.

We negotiated another tuk-tuk ride back to Haad Rin in the pouring rain. Daryl stayed pissed at me the whole way.

When we got back, he didn't want to hang anymore and went straight to his hotel.

Whatever, dude.

It was late and raining anyway.

I had a few more drinks with Austin outside his hostel, then stumbled off into the night myself.

The next morning.

Nobody on the island seemed to be awake yet. The sun was barely peeking over the horizon—just a sliver of gold.

I climbed into the hammock—my hammock—strung a few feet from the ocean. The fabric was cool from the night air, the sand still damp, the tide lazy and soft.

Out there, with nothing but the whisper of waves and the slow warm breath of the rising sun, it felt like the world wasn't rushing me anymore.

Like life finally stopped shouting long enough for me to hear myself again. Feel again. Be again.

And goddamn... I had missed that feeling.

And as I swayed there, staring out into that endless blue, something happened.

For the first time in a long, long time—I felt peace.

Real peace.

The kind that sneaks into your bones before you even realize it's there. The kind that fills up the cracks in your chest without asking permission. The kind that doesn't give a shit how long you've been searching or how far you've dragged your tired soul across the world to find it—when it's ready, it'll find you.

I shed tears for this long-ago friend I never knew I'd see again—and when I finally did, I didn't realize how much I'd missed her.

The tears came easy. Like the body recognized something the mind had forgotten. A homecoming I didn't know I'd been aching for.

I whispered, "Thank you, my friend. I will never take you for granted again. I will hold your hand for the rest of my life and protect you at all costs. In my soul you will forever remain."

Then a question popped up—clean and sharp:

How do I come back here as often and as long as possible?

And as they do when you ask the right questions, the answer arrived immediately, like it had been patiently waiting for years:

Go home. Pick the highest-paying job that gives you the most freedom. Save every dollar. Strip away everything else. Come back here as often as possible to this peace.

That meant downgrading from my overpriced one-bedroom apartment on the swanky side of town—where all I did was come home late and sleep anyway.

I'd been worrying about the stuff in that place for months. Stuff I didn't need. Stuff that messed with my peace. The fact that I was worrying about it pissed me off.

Worry steals peace. Even a second of it.

So I decided right there on the hammock:

When I got back, I'd toss everything.

The 85-inch TV that swallowed the living room wall? Gone. All the pots and pans I don't use? Gone. All the clothes I don't wear? Gone. All the furniture I don't touch? Gone. All the appliances? Gone. All the boxes in the garage? Gone.

After the lease ended, I would live cheap. Minimally. I'd own a single pot, a single pan. Downgrade from a fancy remote-controlled queen to a small twin bed.

Keep only the rice cooker, the microwave, a handful of essential clothes, my small sturdy four-wheeled luggage—and a nice sleek bidet.

I would not be attached to anything or anyone ever again.

All I wanted was the freedom to drop everything at any moment, buy a one-way ticket, and be back here in a couple of days.

For a first step, the idea was pretty damn good.

It wasn't the end goal—but it was the clearest sense of direction I'd felt in years... maybe decades.

I'd been unsure of everything for so long I'd forgotten what clarity even felt like.

I promised myself I'd ride that train until it took me to whatever it was I'd been searching for—wherever it led, however far it took, however many noods I had to slurp along the way.

Honestly, I didn't think I'd find peace so quickly.

I thought I'd have to wander all of Thailand for months—maybe years—like Patrick did.

And I was ready for all of it.

But no.

Somehow it found me after just three stops in Thailand.

But now it was time to go home and clean up my life.

Son of a bitch.

But not until after the party.

I took a dip in the ocean, checked out of that great little hotel, flirted with my favorite Thai front desk girls, and grabbed a tuk-tuk to my expensive hotel up on the hill.

I checked in.

It was owned and run by a lovely Israeli couple with a cute little two-year-old boy who was always laughing and running around. He'd stare at me curiously, but every time I put my hands out, he'd sprint back to his mom crying.

Fair enough, little guy.

Don't ever change. Be careful with people. And most definitely don't marry the first girl you ever kiss.

Their unit sat at the bottom of the steep hill so they could see who came and went. My room was the highest one all the way up—great view, brutal climb. They also had a few camping tents off to the side for backpackers.

I dropped my stuff off and went hunting for a bowl of noods down the street closer to town, waving goodbye as I passed their open doorway like I'd lived there for years.

I slurped down a simple bowl, then walked over to meet Daryl at the charming hotel I'd checked out of earlier.

Austin was already there.

They were both in the infinity pool—Austin leaning over the edge with a Chang in hand, looking at the ocean like he owned it, and Daryl, of course, was already chatting up a girl from his country.

I was happy he'd somehow—however impossible it seemed—recovered from the devastating heartbreak of last night.

Austin shrugged and waved the bartender over.

I slid into the pool next to him, and we drank Changs overlooking the water, hanging over the infinity edge like kings of the coastline.

Hours went by like minutes—only stopping to take a piss, smoke a cigarette, or order another beer.

Knowing we were down a man, we headed out to see how they were setting up for the Full Moon madness the next day.

Changs in hand, of course.

We wandered, explored, and soaked in every bit of it.

The crowd was growing by the minute—people rolling in from every possible ferry, speedboat, private boat, even long-tail boats spilling and sputtering in from God-knows-where.

The noise rose with them: chatter, laughter, music blasting from hidden corners and open hostels.

The narrow streets got tighter. We had to weave around groups forming outside hostels—drinking, buzzing, vibrating with anticipation.

Restaurants filled up. Lines snaked out onto the road. Shops burst alive with noise and neon.

The whole island shifted—from quiet anticipation to full-on chaotic energy.

And it was only the night before.

Berlin chirped in the text that she'd arrived.

We met up with her and her friend—who looked a hell of a lot better than back on Koh Samui but still had the look of death on her. Poor girl.

We sat with them and ate Indian food. Berlin asked where Daryl was.

Austin and I looked at each other, shrugged, and said together: "Don't know."

She rolled her eyes. She already knew exactly what that meant.

"Whatever," she said.

We asked if they wanted to explore some bars with us. Her friend still wasn't feeling well. Berlin said she was down—but only if Daryl wasn't coming.

We told her she wouldn't have to worry about that.

So the three of us went out and bar-hopped the rest of the night—three parts of the original four, back at it again.

We tore through every bar Austin and I hadn't hit yet, and some we liked, mapping out the terrain like spies on a reconnaissance mission for tomorrow's destruction.

Then the sky cracked open.

It started pouring so hard the rain damn near knocked me over. The wind tore through chairs, umbrellas, decorations, lights—sent them skidding and tumbling like paper toys.

Waves crashed higher and higher on the beach, dangerously close to the bars and shops, some rolling up and over the little concrete seawall.

Thunder rumbled behind the palms like a warning from the gods.

If this shit kept up tomorrow, the Full Moon Party would be a funeral.

We all went quiet for a second and sent up a prayer to the heavens.

That was our cue.

We called it early, hugged it out, and saved whatever juice we had left for the real party.

They were staying close by, lucky bastards.

I still had to stumble all the way back north to my hotel through the storm.

I was drenched from head to toe.

"Drenched" doesn't even do it justice.

I could barely see two feet in front of me.

Now I know what it feels like to get sprayed in the face with wet bullets.

I don't think I've been in rain like that since I was a kid.

And somehow, in the middle of that chaos, it was incredibly refreshing.

But the storm was only outside.

Inside...

Inside was this strange, quiet calm I hadn't felt in years.

Maybe—just maybe—the slightest bit of something dangerously close to...

happiness?

Let's not get too greedy, Cali.

Chapter 9: Full Moon Party

When I woke up the morning of the Full Moon Party, it was already past ten a.m. I hadn't slept in that late in years. Maybe I'd finally found peace and my body accepted the invitation. Or maybe it was pure exhaustion wrapped in denial.

It didn't matter. I wasn't going to overthink it.

I lay there longer than I should've—staring at the plain white ceiling, the white walls, the brown blinds. I know color better than most people on earth, but I'd never really looked at a hotel room before. Never studied the cloud-white walls, the alabaster aircon wedged into a hole, the way the light softened everything.

It was comforting. Clean. Quiet. A strange, welcome change.

Since Austin told me about the Full Moon Party, it had existed in my mind as this epic, mythical climax. But lying there like a man without urgency, I realized I didn't need it. Everything I'd come searching for was already unfolding.

The party wasn't salvation. It was icing.

Austin finally texted. He was awake and jittery with excitement—his friend was arriving soon. He wanted to meet up. I told him I'd shower, eat, and wander over.

I walked to a convenience store—one of the rare ones without a 7-Eleven sign—and tried withdrawing cash for the night. The ATM was empty. Wiped out. Probably by farangs—foreign assholes who drain half the country's ATMs dry. I'd only recently learned the word farang and immediately hated farangs in principle.

I crossed the road and ordered a cheap plate of pad krapow from an old grandma. Simple, spicy, perfect. Austin would've loved it. Noods weren't ready yet.

You can't miss today, I told myself, smiling like a fool.

Everything was hitting.

I was walking—strutting—like I'm "feeling alive, feeling alive, ha ha ha, feeling alive," or whatever that song says.

After lunch, I meandered through town to soak in the anticipation. Traffic had transformed from dead quiet to chaotic and buzzing. I visited my favorite front desk girls at my old hotel—one married, one definitely going to the party. I grabbed the younger, unmarried one's WhatsApp and saved her under the most logical name: Front Desk.

Finally, I made it to Austin's hostel with two Changs from 7-Eleven. "Hostel" was generous. The place looked like a human filing cabinet. Beds stacked on beds with thin red curtains acting like imaginary walls. Men on the right. Women on the left. People snoring at two in the afternoon behind their itty-bitty curtains like the world wasn't about to end tonight.

We sat outside while he took a call from his friend. I waited, and that's when this tall, gray, skinny, wrinkly wanker wandered up with a cigarette hanging off the corner of his mouth.

"All good, mate?" he asked.

"Hell yeah, matey. And who the hell are you?"

He laughed. "I own this place. I'm Danny."

"Oh shit. My buddy's snoring right there." I nodded toward one of the coffin-like cubbies.

Danny cracked up. I asked how he ended up owning a hostel on Haad Rin. He rolled into his story like he'd told it a thousand times and still loved it: proper British family, proper life, proper mistakes. Had to run—didn't explain why. Didn't need to. Ended up in Thailand. Stumbled onto the island. Never left. Married the jewelry lady across the alley, who waved at me violently. I bowed back like a decent immigrant boy.

He said she loaned him the money to buy the hostel—ten grand he didn't have.

"Couldn't pay her back," he shrugged, "so I married her. Best mistake I ever made."

"Well, Danny Boy," I told him, "bumping into your sorry ass might be mine."

I gave him my sad story. He smiled and said, "You're more than fine, mate."

And just like that, we became brothers for half an hour—laughing, hugging, trading the kind of fleeting confessions only strangers in paradise can afford.

Another possible path unlocked.

Meanwhile, backpackers from every corner of the planet spilled into the alleys. Tattoos, accents, hairstyles I'd never seen before. This island was becoming a living organism.

Austin reappeared. Said his friend was arriving in a couple hours.

We stopped at a bucket stand run by a semi-celebrity island character named Alex. Shirtless. Glow paint splattered across his chest reading: ALEX IS A GOOD CUNT. A huge sign above him said: ALEX, FUCK MY BUCKET. You don't meet many people like that and forget them. He knew the toilet lady running the pay-to-pee stall next door. Because I knew better, I gave him 500 baht up front for unlimited bathroom access—for me and the boys and girls later.

Insurance.

He was thrilled. So was I.

The streets were swelling with bodies and noise. Vendors scrambling. Bass thumping.

It was time for a nap before the storm hit. Austin agreed.

I bought a flimsy neon yellow "Full Moon Party" tank top and a tube of glow paint. I wasn't trying to be cheap—I just didn't have time for the glow-in-the-dark artists to tat me up. I walked past the pier where boats were unloading bodies like a zombie apocalypse, set my alarm for 8 p.m., and knocked out.

I woke from a dream where I was already having the time of my life.

Wonderful.

A great party in my dreams, and now I'm about to go have a great party again.

A great omen.

I showered, beaming the whole time. Painted myself poorly—glow paint everywhere. Rookie mistake, especially on the face. Sweat melted it off instantly and got on everyone and everything.

As I passed the owner's unit, he waved me over.

"Let me take you to the party, brother!"

I hopped on the back of his scooter. New me accepts gifts. New me remembers not every hand extended wants something in return.

He gave me three warnings as he dropped me off:

1. Don't drink the buckets—sugar and alcohol hit harder here.
2. Have a meeting spot—you'll all get lost at some point.
3. Don't jump the flaming rope—you'll get burned.

I proceeded to ignore all of it.

We met up at Austin's hostel. Finally met his friend Seattle and her soft-spoken brother. Berlin arrived—no sign of her sick friend. Daniel and his gorgeous sister from Canada appeared out of nowhere. Apparently, Austin met them while I was napping. Then Front Desk and her friend—Massage Girl—wandered by, so we all joined up.

Our group went from three to nine before I could even process it.

I immediately bought a bucket for the guys.

The girls were smart.

We held hands and pushed into the glowing, sweaty madness.

Within minutes, Seattle and her brother were gone forever. No signal. No hope.

The beach was packed with fifty thousand people. Rain poured intermittently, which somehow made it more romantic. Thunder rolled off and on, and we danced harder when both hit at once. The rain cooled our skin as neon paint streaked down our faces.

It was cinematic.

Tears in rain.

A baptism.

We danced and screamed through every inside and outside bar. Danced with strangers from every place imaginable. All here for the most epic, craziest, happiest place on earth. Every type of music blasted all at once, but wherever you were, you were perfectly in tune with only that one sound.

I was screaming "Da me mas gasalina" by Daddy Yankee when it hit. Thousands of us were chanting "This is how we do" by The Game and 50 when it came on. Of course, you can't forget when everyone automatically performed "YMCA" together in perfect unison. All songs mixed and mastered to the flawless mood on that beach.

Strangers from all over the world were brothers, sisters, and lovers.

Nobody asked who you were back home. Nobody cared what your job was, what your pain was, what your divorce papers said, or what your bank account looked like.

Out there, the ocean didn't recognize titles.

The moon didn't recognize trauma.

The beach didn't give a shit about loneliness.

All that mattered was that we were alive. All that mattered was that the night was still young, the bass still thumping, and the sky looked like it had been painted just for us.

And God—the colors.

Neon paint smeared on cheeks and bodies like war paint. Glow sticks snapped like fireflies being born in our hands. Flames roared all around. Buckets sloshed. Cigarettes glowed like tiny lighthouses. Salt in the air. Sweat in our eyes. Laughter everywhere, coming from places you couldn't even see yet.

It felt like the world had ended and restarted on that beach—and somehow every broken soul in the world got the memo.

You'd stumble into one bar and it was pure chaos—techno screaming, people bouncing like they were possessed, the floor vibrating like a heartbeat. Then you'd spill outside into the sand and the beach would swallow you whole. Waves crashing, feet sinking, music changing instantly like the universe just switched channels.

And somehow it all made sense.

Somehow it all worked.

It wasn't one party.

It was a thousand parties stitched together by alcohol, moonlight, and desperation.

And you'd just drift through it like a drunk planet with no orbit—pulled by whatever sound hit your chest the hardest.

One moment you're dancing with a Swedish girl and a British guy and some tattooed Aussie who keeps yelling "LET'S GOOOOO!"

Next moment you're hugging a stranger from Germany like he's your cousin you haven't seen since childhood.

Then you're screaming lyrics with people you've never met, louder than you've screamed anything in your whole damn life.

And it wasn't fake. It wasn't pretend friendship. It wasn't shallow connection.

It was real, because it was temporary.

We all knew none of it would last. We all knew we'd disappear back into different countries, different time zones, different lives.

But for those few hours?

For that one full moon?

We belonged to each other.

Then a couple girls needed a bathroom. I dragged them to Alex. Fist-bumped him. Free entry. Another bucket. Then Daniel and his sister vanished—an emergency. A friend was having an episode or something. I got her number but somehow lost it.

Not asking Daniel for it.

Man rules.

Then—because I'm a bonehead—I decided I was a jump-rope champion and hurled myself into the flaming rope like an Olympic gymnast. I thought I did magnificently. I got torched. A couple burns. Crowd cheering.

It was not magnificent at all.

And then, suddenly, I lost everyone. No phone. No plan. No meeting point.

Just rage and regret, stumbling alone through neon chaos.

I walked back to Austin's hostel hoping someone would appear. Nobody.

Then—miracle—Front Desk and Massage Girl passed by.

"Take me home!"

After confused and slurred translations, yelling over the music, and what I can only assume was divine intervention, they understood.

After that, it was just black for me.

Somehow, I made it up the steep hill to my hotel, crawled into bed, and crashed.

The next morning, Austin called around noon.

"You were fucked up, bro. Meet at Daryl's."

We met at my favorite hotel—Changs in hand. Burmese cigarettes rolled by the bartender himself, a soft-spoken guy who told me he dreamed of seeing America one day.

I said, "If you ever make it over, I'll show you all the America you want to see."

"Deal," he said, smiling wide and honest.

We added each other on WhatsApp like two kids making a pact under a coconut tree.

Austin caught me up. Evidently, at some point in the night, I handed him my phone and begged him to record my drunk ass attempting a cartwheel in the ocean—if you could even call it that. In the video, I do a little twirl and fall straight on my face while Berlin cackles loud enough to drown out the music.

Then he said, when they turned around—I was gone.

They searched everywhere—through crowds, down alleys, by the bars, along the waterline—no luck. After making sure Berlin was safe, he decided I must've wandered back to my place. He remembered my constant bitching about how overpriced my hotel was and stormed up the hill to find it. He marched straight to the owner's unit at two in the morning and demanded he open my room to check if his "gay little Asian friend" was alive.

I was.

Not gay, but face-down, dead asleep, fully clothed, covered in glowing paint.

He said he tossed my phone onto the bed beside me, shook his head in disgust, and left.

The fucking buckets, man.

Daryl finally reappeared—the long-lost brother—emerging from his hotel room smiling like a man who'd just survived a battle.

We spent the afternoon polishing off the last of the Burmese cigarettes and a few more Changs. I flirted with Front Desk, then wandered down the street for a massage from Massage Girl.

The girls filled me in on the gaps from the night before—they found me, figured out where I was staying after several very bad attempts on my part, and eventually dropped my stupid ass off. Apparently, I also proposed to both of them and promised we'd all live together in one big happy house.

Sounds about right.

We watched ferries carry away waves of hungover zombies.

Then the Three Amigos—me, Austin, and Daryl—shared our last dinner together. A simple Italian meal to give our tongues a break from all the Thai we'd been devouring. I had a calzone—all meat. Laughs. Hugs. The kind of goodbye you only earn through shared miseries and memories.

Austin was heading off with Seattle to the jungles of Khao Sok National Park. Daryl was flying to Bali.

And I was heading home—to dismantle my life piece by piece, purge the weight holding me back, and clear the runway for whatever the hell I was meant to chase next.

Chapter 10: The Long Way Home

Koh Phangan was another planet now.

By sunrise, the thunderous roar of the party crumbled into a distant hum. Cleanup crews were just beginning their second slow march to the scene of the crime. Most of the litter was gone, but plastic cups were still everywhere. Empty bottles clustered like small memorials. Beach chairs sat toppled and scattered, as if some strange plague had torn through in the night. Ownerless flip-flops dotted the shoreline like casualties. Neon paint streaked the palms—most of it mine.

Even the ocean looked exhausted.

I knew I wasn't leaving with the rest of the wounded soldiers. I'd already seen the lines—tens of thousands of partiers limping toward ferries and speedboats like a biblical exodus. Hollow-eyed. Voiceless. Sand glued to their hair. Regret steaming off them like heat waves.

God knows I had some of my own.

But I wasn't going out like that.

So I stayed. Took my time. Moved slow.

It's easy to linger on a tropical island you never want to leave— especially when you know reality is waiting somewhere out there, sharpening its hand, ready to bitch-slap you back to earth.

I packed up quietly, said goodbye to my little Israeli family—they hugged me like a son—and went for one last bowl of spicy grandma noods. The best kind. The kind that burns your lips and your regrets at the same time.

The story of my life.

At the pier, I bought a ticket for a speedboat back to Samui. No line now. Just stragglers and hangovers. Everyone boarded silently, like we were attending a funeral for all our dopamine. We bounced across the water, heads down, barely a word among us.

A giant ocean-wide serotonin crash.

It was the saddest scene I'd ever witnessed.

A taxi. A flight. A blur.

Then—Bangkok again.

I thought about heading downtown. Saying hi to my favorite bartenders. The street vendors who smiled when I walked by. The people who remembered my Chang order without asking. But the thought of the traffic alone made me want to vomit. They missed me one day. Maybe two. Maybe forever. I am, according to Daryl, a goddamn unicorn.

So instead, I found a cheap room north of Suvarnabhumi—quiet, simple, perfectly normal.

That night, I went hunting for noods. I followed the smell of broth down a dim street until I found a tiny shop with wobbly plastic tables on the sidewalk. No English signs. Just a couple locals and a big ass pot that looked older than Thailand itself.

I ordered a bowl of beef noodles, spiced it until I was sweating, and took the first slurp.

And right then—right in that moment—something snapped into perfect clarity.

Anthony Bourdain, my guy, was right.

All the things I need for happiness:

Low plastic stool? Check.
Tiny plastic table? Check.
Something delicious in a bowl? Check.

I didn't need therapy. Didn't need a five-year plan. Didn't need advice. All I needed was to sit on a dirty Bangkok street with a bowl of noods.

And in that quiet moment—mid-slurp, sweat dripping down my face—I knew exactly what to do next.

I chose the job I was meant to take. Not the one with the bigger salary. Not the one with the fat corporate title. Not the one that would chain me to only two weeks of vacation a year..

The job that bought me freedom.

Sales and marketing for a contractor friend. A role perfectly built for someone like me—someone who could talk to anyone, read a room, build trust, and make shit happen.

I texted the owner—one of my biggest accounts back at the old paint company—that I was coming home. Then I followed it with a call.

"Yo," I said. "I'm coming back soon. Have an offer letter ready. Salary, commission, a truck. And every winter when construction slows, I'm gone for a month. Thailand. Every year. No exceptions."

He laughed. Then he agreed. No problem. Because he knew what I could do for him—like I already had from the beginning of his company.

In that moment, with nothing but a bowl of noods and a plastic stool under my ass, I found the next step.

Sometimes that's all life gives you—just the next stupid step—and that's enough.

The next morning, I packed everything—tank tops, shorts, twenty colorful shirts stuffed into my bag—and headed to the airport. The check-in woman said my bag was overweight or too big or something. I had to check it. Like an amateur, I just handed it over.

Nothing could ruin my mood.

Then came the International Departures escalator. Every Instagram influencer calls it *"the saddest escalator in the world."*

For once, Instagram told the truth.

I floated through security like a man reborn.

My layover was in Hong Kong.

A childhood dream.

In my head, I imagined stepping off the plane into a live-action kung-fu movie. Slick neon. Badass poses. A dramatic zoom-in on my face. Bruce Lee. Jackie Chan.

Instead—I landed at 1 a.m.

It was cold. Very cold.

And I was wearing:

- a tank top
- shorts
- flip-flops
- an idiotic island hat
- an idiotic fanny pack

Everything else was somewhere in my checked luggage, probably being punted around by a pissed-off baggage handler in another country.

I told the taxi driver, "Take me downtown!"

He did. Dropped me in the middle of Kowloon at 2 a.m. I was the only moron dressed for the beach in a city deep in winter.

It hit me instantly:

I had severely, catastrophically fucked up.

My flight home wasn't until 5 p.m.

Fifteen hours away.

No hotel. No jacket. No plan. Just an underprepared piece of shit shivering on a sidewalk in China.

No use beating myself up though.

I had a secret weapon.

Jameson. Double. And keep them coming.

I spent the next several hours barhopping through Kowloon—young locals staring at me in disgust, confusion, or fear. I looked like a lost beach uncle who teleported into the wrong movie.

The tannest motherfucker in all of China.

When the bars closed around four or five in the morning, I met a Thai couple—my guardian angels. One look at me and they knew.

"You just came from Thailand, right?"

They took pity on my frozen dumb ass. Bought me a drink. The guy handed me his XL Nike jacket, which instantly became the most precious object I owned.

I tried several hotels. All full. All unwilling to even let me sit and warm up in the lobby. I didn't want to end up in a Chinese prison for trespassing or "existing weirdly," which I was most definitely doing.

Finally, after wandering like a drunk ghost, I stumbled upon the savior of all saviors—Kowloon City Park. There were stairs. I climbed. There was a bench. I collapsed.

And fell asleep wrapped in the XL jacket of a stranger.

I woke to traffic, horns, and a small herd of cats sitting on me like I was a space heater. One perched on my stomach. Another curled behind my knees. Another sniffing and licking my face.

Probably gave me seventeen diseases.

Didn't care.

They saved my ass from becoming an icicle.

The sun was rising and Hong Kong was exploding awake. Markets opening. Buses roaring. Neon buzzing back to life. I stretched, cracked my back, and thought:

"Well... that could've been real bad."

Then I ate my way across the city—noods, dumplings, chicken feet, skewers, mystery seafood, and anything better left unquestioned. I walked until my feet bled. Figured out the train system and rode it everywhere, snapping photos and videos like proof I'd actually been alive that day. Washed it all down with roughly ten Heinekens—perfectly calibrated to the gray, overcast mood hanging over the city.

Eventually, exhausted and about forty percent feral, I returned to the airport. Shuffled through security. Napped against a cold wall at my gate.

Then I finally boarded the flight home.

And for the first time—

…I slept on a plane.

Deep. Heavy. Dreamless. Peaceful.

As if the universe finally said:

You've earned this one, Cali.

Chapter 11: Back to the Grind

As soon as I landed—on a Tuesday—I got sick and caught the worst jet lag of my life. Thailand is about fifteen hours ahead, so my body stayed on island time out of spite. I'd be wide awake all night, coughing and eating like a nocturnal raccoon, then drag myself into work all day like a zombie wearing human skin. I'd crash hard around 10 a.m.—which is 1 a.m. in Thailand—and the absolute worst hour was 1 p.m. here, the equivalent of 4 a.m. over there, when my body insisted I should be dead asleep.

I lost my voice. I was sick, tired, permanently tilted. Most days I had to sneak in two-hour lunch naps just to survive. I wasn't living—I was enduring. Only sheer stubbornness—and the fact that I actually loved the job—kept me from falling apart.

Then Friday came and my body finally mutinied. I wiped out at 7 p.m. and didn't wake up until 7 p.m. the next day. For a full minute I thought it was morning.

Then it hit me:

I'd slept through an entire day.

Jet lag had kicked my ass. Then the cold finished the job.

I was good at what I did because I genuinely loved people. It came naturally. I listened. I paid attention. I asked real questions. I never gave off used-car-salesman energy—I wasn't trying to close anyone, I was trying to help them. People felt that. They trusted it. That's how I became the best rep in the country—the highest-paid rep in company history.

I always gave people the benefit of the doubt. I did whatever I could to help them succeed. I chose—deliberately—to see the best in them. But what I never did was look hard enough at the worst.

I ignored warning signs with my ex. With so-called friends. With contractors who took advantage. With companies that would sell you out in a heartbeat. I kept choosing optimism over reality. Hope over evidence.

Mom always warned me. "That's your greatest weakness, Son," she'd say. "You love and trust people too much. You've helped so many without asking for anything in return. Maybe I raised you too well. Maybe that's why you ended up with her."

I never learn.

That's how I ended up with these wolves too.

Work welcomed me back with familiar faces—old friends in a new world. I'd been their paint sales rep in what felt like another lifetime. Now I was their Sales & Marketing Executive. Their bridge to something bigger. For a moment, it almost felt like destiny.

But like any marriage, the honeymoon is the best part. And once it ended, the ugly truths started walking around the house without clothes on.

We were a painting company pretending to be a construction company. A painting company run by men who wanted the title of ownership, not the responsibility. Ego everywhere. Commitment nowhere. They carried old wiring—violent pride, quick tempers, that constant need to dominate the room.

And I was supposed to sell it.

The ownership group used to be four, then three, then finally one after the others got sick of the lies.

I knew paint better than almost anyone on this planet—but paint is the last step. The easy step. At the time, I didn't understand everything that came before it. And the "construction expert" who was supposed to teach me? All he fed me was lies dressed up as confidence.

The first job I sold was small: a hole in a bathroom ceiling. Drywall, texture, paint. Easy. He told me it would take two guys three days. So I wrote the proposal for $3,900 and sold it. A friend at a management company pushed it through. The owner signed.

The truth? It took one guy half a day to patch and texture, and half a day to paint.

When I said we should credit the client, he laughed in my face. "Sucks for them," he said. "They're suckers. We just made 800% profit." He said it like he'd discovered math.

I threw up in my mouth.

Another time he told me replacing concrete steps at an old property would cost $5,500 each. Fifty steps. I questioned it. He launched into an Oscar-worthy performance—huffing, puffing, inventing complexity: "You gotta remove this... shore that... weld this... pour that... three guys, three days per step..."

And like a fool, I believed him.

I sold the job.

Turns out the steps weren't custom at all—they were sitting on pallets in some yard, rotting. The real cost was a couple hundred bucks a step. I felt so guilty I did a ton of extra work for free just to make it feel less wrong.

That's when it hit me:

This "expert" was the biggest pretender of all.

And I learned something ugly about prolific liars—the kind I wish I didn't know: They're not good at lying because they remember everything. They're good at lying because the second it leaves their mouth... they believe it. Otherwise they'd have to track the damage. They'd have to see other people as human. But to them, people aren't people.

They're prey.

He was also the owner's best friend. And the owner? Same species. Just a more evolved kind of rot.

The "expert" started getting threatened by me—by my results, by my relationships, by the fact that I wasn't like them. So his mission became simple: gather power, gather resources, then use them to make me look bad. He'd send his guys to job sites to hunt for faults, exaggerate them, or just invent them—then report back like it was scripture.

Eventually I snapped. One day he lied straight to a customer, and I finally told the owner I quit.

The owner chose me over him.

On his way out, the worst thing he could say about me was that I was a prima donna… and my desk was messy. That's all he had.

Pathetic.

So I spent nights and weekends on YouTube learning construction and reading plans, because if I didn't teach myself, nobody was going to teach me. But the hardest lesson wasn't construction.

It was this:

You can know someone for years. Call them a customer. Even call them a friend. But until you work with someone day after day—late nights, weekends, stress, money, power—you don't really know who they are.

I ended up surrounded by the kind of men who were rotten in ways that spread like black mold. Cheating. Lying. Stealing. Screaming. Pretending. Everyone circling each other like wolves, then calling it "business."

The best part of the job was what I was actually hired to do—customers, marketing, relationships. The stuff I loved. The stuff I was built for. But someone had to handle the numbers. And somehow, that somebody became me.

So there I was—hunched over a laptop night after night, estimating work I never wanted, becoming good at it purely out of necessity, while the "construction guys" played salesman even though they hated people and had no idea what they were doing.

The only thing they were good at was passing around the receptionist.

They'd meet clients, make empty promises, lie about timelines, forget what they said, then end their nights at strip clubs blowing company money. Meanwhile, I was the first one in and the last one out—working past 10 p.m. most nights. I'd get home exhausted, worried, wired, unable to sleep.

So I drank half a bottle of Jameson—my medicine—and knocked out.

At one point I counted it: forty days straight. Almost sixteen hours a day.

Then the company started bleeding out. Payroll bounced. Cash disappeared. The office doors would get locked while employees banged outside—because the trucks were parked out front and everyone knew we were inside.

One time payroll couldn't be made, so I loaned him the remaining $30,000—stupidly, without a written agreement. I did it because I didn't want the company to fail. I did it because I cared about the employees. Nobody put a gun to my head.

But it should've meant something.

When money came in, did he pay me back?

No.

He went to Vegas and did what men like him always do—blew twenty grand on hookers and cocaine—then crawled back like nothing happened. The bookkeeper showed me the details. He was a VIP at the strip club for a reason. He lived there.

And the moment everything finally clicked didn't come from gossip or office drama. It came the day I stopped being naïve and looked him up the way you look up a used car before you buy it. Court records. News articles. A paper trail.

I'm not here to list charges or play lawyer, and I'm not turning this book into true crime. But what I found wasn't rumor. It wasn't exaggeration. It was public record—enough to explain exactly what kind of corruption I'd been standing in every day.

When I finally read it, it didn't just disappoint me—it split something in me clean down the middle.

And the sickest part?

In private, he wore it like a crown. Like being a monster was some kind of achievement. He even asked me to write a book about the vile, predatory things he'd done—as if infamy was better than anonymity, as if being hated was better than being forgotten.

And the longer I stayed, the worse my back got. Cold weather shredded me. Stress was one thing. But the soul? That's another. I didn't belong there—not with those people, not for any amount of money. My body knew it and screamed every night.

And the worst part?

They still owed me money. Every time I asked, the answer was always the same: "Next week."

I dreaded that answer. I hoped for a different one. I couldn't breathe anymore. And slowly, I stopped being able to think at all. The peace I'd promised myself in Thailand—peace I swore I'd protect—was almost gone.

So I started repeating the only prayer I had left:

Just get out.

Just a couple more months until Thailand.

One mountain a day.

You got this, Cali.

Chapter 12: The Dating Comedy Show

In the middle of the worst job I'd ever had, I figured I'd try dating.

I swore I'd never touch a dating app, but my buddy G-Money told me to try Facebook Dating because "it's not really a dating app". Supposedly it was more like friends-of-friends meeting friends. Whatever that means. I wasn't meeting single women at the grocery store or at those quiet, overpriced dive bars full of retired millionaires near my one-bedroom apartment, so I figured I had nothing to lose.

I signed up. Let Google help me write a bio. Huge mistake.

It read like a woman wrote it for women who don't exist in real life. Kind. Funny. Outdoorsy. Loves working out. Long walks at sunset. Loves animals. Absolute nonsense.

Then came the photos. I had to dig back years to find anything usable that wasn't me taking terrible golf course selfies with the boys. I cropped most of them. Even the dog photo was from the damn golf course. Cropped that too. But hey—I worked with what I had. And Jesus Christ, it was exhausting.

Dating apps are just job hunting with worse odds. If you want a job, job hunting is your job. You don't send three resumes a month and hope the universe rewards you. You fire off hundreds. Dating's the same thing, but more cutthroat and way less honest.

I swiped through everything within 25 miles. Then 50. Then 100. Eventually I started seeing the same profiles over and over, which is how you know you've reached the bottom of the algorithm. I didn't care enough to learn how it worked.

But something did stick out. So many profiles said the same thing: "Don't bother if you're divorced less than two years."

Mom had told me the same thing after my life blew up.

"Son, I know it feels impossible right now. Like you can't breathe. But time helps. I've seen it a hundred times. It takes about two years to get over a divorce."

She listed cousins and aunties. Family I didn't remember and didn't care about while in my current state. But the message stuck.

I went on a couple dates anyway. They looked nothing like their pictures. Shocker.

I gave up on the app.

The women were right about one thing though—honestly, I wasn't ready. It had only been a little over a year. I needed at least two. Still, I reasoned that I needed practice. I wanted to be ready when the time came. I didn't want to limp onto the start line. I wanted to restart my life sprinting.

God kept laughing.

No more apps. Real life instead. Old-school. Awkward. Painful. A numbers game. No time like the present. Fresh back from Thailand. Slightly hopeful. Completely out of practice.

The receptionist at work mentioned her sister didn't have a Valentine's date. Why not?

I dressed up for the first time in years. Took her to dinner. Taught her Pai Gow — the best drinking card game in a casino. She won some money. We split it because I fronted the cash. That was the deal. The night felt good.

But it was late. Tuesday night. We both had work early. I didn't know how to close. So we stood there awkwardly like two people waiting for instructions.

"Another time?"
"Friday?"
"Yeah. Friday."

Out of practice, man.

The next day the receptionist said her sister loved the date. Was excited.

Friday came. I got stood up. Food was good though.

Monday, I asked if her sister was okay.

"Oh yeah," the receptionist said casually. "One of her boyfriends found out."

What the fuck. Deleted her number immediately.

Some time later, I was being loud at one of my favorite Mexican spots — liquor-loud, cracking jokes. A skinny, blonde sat a couple of seats to my right and laughed at everything I said. Eventually – with enough liquid courage – I walked over.

"Why are you laughing at me?"
"Because you're funny."
"Well shit. We already have one thing in common."

Eyebrow raised.

"You like me. I like you. That's two. We both drink. That's three. Want to take a shot?"

She smiled. We took Don Julio reposado. She made the yucky face. I respected it.

"You better just give me your number, this list is getting long."

Got her number.

We danced. My back hated me. Shirt came off. It always does. My signature move.

We kissed.

She was an incredible kisser. Like—holy shit.

She said I was too. I don't believe her. I've only heard that once, and it was from her. I've kissed exactly two women in my life now. My sample size is weak.

She wanted to take me home.

Cab ride. Heavy making out. Driver missed a red light in front of the police station. Lights immediately. Cop walks up. I try to apologize. She tells me to shut up. The driver says he just got out of prison and this is his only job.

Cop looks at me. I give him a thumbs-up.

He hands the ID back.

"Be careful."

We make it to her place. Wine. Jameson. Couch. Clothes coming off. I tasted every inch of her. Then she went down on me.

And I froze.

I pushed her away.

"Sorry, I just got out of a 25-year marriage," I said. "I promised myself I wouldn't do anything stupid."

You're overthinking it, she tried again.

I pushed her away again.

"Last time I didn't think, I ended up married and divorced."

She shrugged. Opened wine.

I asked if I could just kiss her all night. She said yes. We did.

It was one of the goddamn roughest nights of my life.

Friends said I was definitely overthinking it. I eventually gave in.

Three weeks of light in a very dark time. Then she ghosted me.

Bestie said I probably dodged a bullet. Single mom. Three kids. Serial dater. Sugar daddy energy.

A year later I saw her again. She wanted a hug. I walked right past her.

San Diego came next. Old college crush reappeared on Facebook. Emojis. Likes. Messages. She was mid-divorce. She invited me to a wedding in San Diego. Bought my ticket. Airbnb. Couch arrangement.

I thought about it. Nothing was there. Too many old memories of being passed over when I had nothing. Now suddenly I'm interesting? Nope. And she was in free-fall. I wasn't going to take advantage of that. So I let it go.

She chased. I ran.

Chicago sealed it. Phoenix fling. Flight attendant. Should've known. Flew to Chicago. Amazing city. Until she casually mentioned one drunken night she lived with her 60-year-old boyfriend.

"But don't worry," she said. "He won't mind a threesome."

End of story.

I swore off dating entirely. I joined Group Therapy instead.

We were all in the same boat—fresh out of toxic relationships or still clawing our way out of them. Everything raw. Everything burning. So we huddled together and tried to soften the blows for one another the only way we knew how: dark humor and endless reels.

The group chat never slept.

"Rise and fucking shine, you Twat Waffles!"
"Spread the fuckery!"
"You got this, you Sassy Ass Clementines!"

It was ridiculous. It was fitting.

They were all clients of mine back at the now-defunct paint company. We used to bump into each other at events, laugh politely, talk shop. Once they found out I was divorced too—also trying to rebuild from the rubble—they pulled me in without hesitation.

And everyone knows the initiation is simple: you get added to the group chat. That's how you know you matter.

Sweet Tea was the ringleader—soon to become my female best friend. We share the same birthday, which means I never forget hers. I always joke that she'd be perfect for me, but she's got too many kids and too many opinions.

She's the planner. The initiator. The event-finder. Somehow knows every concert, festival, pop-up, and opening within a 200-mile radius before anyone else does. Sends dates, flyers, screenshots, schedules nonstop—like it's her full-time job in between sprinting across state lines for random shows.

She's the heart of the group. Without her, the rest of us wouldn't leave our houses except for work. She's been broken longer than the rest of us—years ahead in the damage timeline—and she knows it. Even has a real therapist. Thoughtful in her own way. Sharp. Protective.

We hang out the most. Always finding excuses. But I never post anything because her ex-husband—who was my friend before I even met her—would lose his mind. Man rules. Lines I don't cross.

Then there's Kitty Kat. She's been trying to escape her live-in ex for years. The cat-lover of the group—those are her babies. I've lost count how many she has. Her reels are mostly cats. Or baby goats.

She's vegan by choice, which makes going out to eat a logistical nightmare. We once went to hibachi in Vegas and everything—even the rice—was cooked in butter. They had to make her something special in the back: a sad little bowl of wilted greens and carrots.

She ate it with a smile. That's who she is.

She gives constantly—especially to homeless kids—because once upon a time, she had no one. Fiercely protective of me. This one would beat a man's ass for me without hesitation. Hardest person in the group to get out of the house. Takes a full-team effort.

And then there's Snooky—the youngest. She got out of her divorce before I did. Her ex beat her, so she finally walked. Two boys—almost grown. One in college, the other close behind. Her kids and her work are everything.

Men line up for her, but she doesn't have the time or patience for anything serious. She's had enough bullshit from men to last a lifetime. When she has her boys, she's not coming out for anything.

She's my wingman. Gotten numbers for me. Almost kicked a girl's ass once for talking stupid to me.

I get a lot of hate for hanging out with them.

Contractors whisper that I only get business because I sleep with them. That I sleep with all the other managers too. It cracks me up that this short, bald, puny Asian guy is seemingly pulling that much ass. News to me.

They're the only people I consistently see outside of work. We find any excuse to get together. They've seen me at my absolute worst—Vegas drunk, chugging half a bottle of tequila on a golf course, ripping off five birdies in a row, attempting questionable flips on camera, then showing up blacked out at the bar but somehow still operational.

They watched me stagger home, force them to get dressed to go out on the Strip while I squeezed myself into stupid leather and see-through lace—only to pout halfway there and announce I didn't want to go anymore.

They were furious at blackout me. Sober me didn't remember a damn thing. Better that way.

And somehow—against all logic—they still loved me afterward.

They're my only real family in Reno. I'd do anything for them.

Otherwise? There's nothing here for me but a lousy paycheck.

And winter was coming. Winter meant work would slow. Winter meant escape. Winter meant Thailand.

It couldn't come fast enough. As soon as it hit, I said: Adios.

Chapter 13: Baptism

I bought a one-way ticket from Reno to Edmonton, Alberta to grab the boys from Canada. Flew them back to the States—to Sacramento—for Christmas with my family. It's always the biggest hassle and the most expensive ordeal when you're not flying out of a major hub. Always the earliest damn flight to either Seattle or Denver, then a three to four-hour layover, then another plane to Edmonton where it's already minus twenty-five degrees.

I'd usually land around 11 p.m., exhausted, frozen, irritated, then fork over an expensive Uber ride to her parents' cramped condo in the far northeast corner of town. I'd end up sleeping on that same small, bumpy, sliding-off-the-frame mattress next to the little one who was complaining about how loud I snored.

With barely an ounce of sleep and dead tired, I'd get them ready the next morning and Uber right back to the airport because I couldn't stand being in that place—those memories—one minute longer than necessary.

Then it was the same routine in reverse.

Thank God the boys are experts at traveling now. It's not herding cats anymore. We've got a system: waiting in line for security, waiting in line for U.S. immigration, waiting in line to board, waiting in line for people to sit their asses down.

Then the connecting flight—to Seattle or Denver only. Those are the principles. Quick two-hour hops, the shortest layovers, and absolutely no connecting through Vancouver because I refuse to make an already miserable trip worse or flying past Reno to SF, LA, Phoenix, or Vegas then back. I refuse. I don't fly over my town just to fly back because it's cheaper. Some things are sacred.

Then it's another late-night landing, usually around 11 p.m. or midnight, then another Uber to a hotel because my shack is too small for the three of us.

Next day: drive across the Sierras—praying it's not snowing—to Sacramento for a week with my brother's family. Mom and Dad come down. The nieces terrorize the boys, as young four and six-year-old nieces do best. After a week, drive back over the mountains—again hoping for no snow—stay one more night in a hotel, then wake up at some ungodly hour and repeat the whole process back to Canada.

Most people would say that's an impossible ask.

For me, what other choice do I have?

The easy choice—the logical, inexpensive, selfish choice—would be to let them stay in Canada and never deal with any of this. They don't like being dragged away from their friends and their gaming computers either. Nobody likes it.

But I remember what a friend once told me:

"They may not like it now, but they'll remember it for the rest of their lives when they're grown."

So I suffer through it. Because that's what good fathers do—they suffer so their kids could have a chance at a better life.

I finally got home—wrecked, half-dead—packed, turned around, and took the first flight out on another one-way ticket.

Next stop: Tokyo.

Flew into San Francisco for the three-hour layover. Ate some halfway tolerable airport ramen to prep myself for what was waiting for me in Tokyo a day later: nood heaven.

This time, I came prepared.

Book about travel — check.
Neck pillow — check.
Light jacket strapped around my waist — check.
Small backpack to double as an emergency pillow — check.
Fancy active noise-cancelling AirPods — check.
iPad Pro loaded with forty hours' worth of Netflix and Amazon Prime downloads — check.
Portable chargers in plastic bags — check.
Brand-new $1,000 4K camera — check.

Professional traveler mode unlocked.

The flight to Tokyo was a cakewalk compared to last year's disaster. Last year, I didn't have a clue what I was doing. This year I was on a goddamn mission.

I purposely booked the flight with the ideal layover—thirteen full hours. Plenty of time to get into the city, cause some damage, and still get enough sleep to survive the next day's seven-hour flight to Thailand.

I arrived excited and refreshed. Breezed through immigration. Checked into a fancy hotel with the expected high-tech bidet situation. Dropped my stuff and took a taxi straight into downtown Tokyo.

Of course, I had to hit the famous Shibuya Crossing and get some Instagram-worthy pics and vids. Then I wandered through every *shotengai* I recognized from YouTube—those centuries-old alleys packed with tiny bars barely big enough for four souls—drinking every Japanese beer, sake', and whiskey I could get my hands on between devouring two of the best bowls of spicy ramen I've ever tasted in my life.

Then I grabbed a taxi back to the hotel—perfectly nooded-out and pleasantly drunk—just in time to pop a couple Advils and pass out with what I'm sure was a stupid smile on my face.

The next morning—after exploring my fancy top-of-the-line bidet—it was Thailand-mode time. The seven-hour flight felt like pure hope and excitement.

And then—finally—I stepped off the plane, walked out of the airport, and my back pain… vanished. Melted away like it had been waiting for this exact moment.

I took a long breath of that familiar warm, sweet, hopeful air—heaven on the tongue, even with the buzzing and screeching of the city all around me. I had to tell myself, calm down, buddy, let's not get too crazy now, okay?

I took a taxi to the hotel and checked in. S31 Hotel. I would grow to love this place. Middle of all the action. Seventy bucks a night. Clean, immaculate, and surprisingly charming—quirky corners, warm lighting. Not a perfect square like most fancy hotels. Heatable bidet. Honestly one of the best, most comfortable hotels I've ever stayed in.

There was a sign on the mini-fridge that said "No Durian." Odd, but makes sense. Durian will stink up the entire floor for three or four days. No exaggeration. The most expensive fruit in all of Asia. Like a mango and a pineapple had a child together and breast fed it with steroids. Huge, delicious, stinky monstrosity. Acquired taste. Asian uncles and aunts can't get enough of it. I'm sure some uncle absolutely tried sneaking durian into the fridge and the hotel now must put those signs up forever.

I dropped my bags and headed out to find the closest bowl of nood. Left or right? Chose right.

Across the street were rows of lively bars and a few glowing street stalls. I crossed over, made a mental note of an empty bar with a handful of girls smiling sweetly as I passed, and kept walking. After about a block I spotted a tiny sidewalk nood shop—three small tables, small plastic stools, random locals hunched happily over their bowls, and a simple metal cart pumping out steam and satisfaction.

Two options: chicken or pork. Chose pork. Fifty baht. A dollar fifty. Sat down and ate what was, at that moment, the best damn bowl of nood I'd ever tasted—simple, savory, spicy. The perfect trifecta. And super cheap, that makes it even tastier.

Girls love me here, of course. Being short doesn't matter—I'm still taller than all of them.

But I'm not here for cheap sex.

It's too easy. Too empty. I've never understood how people meet for five minutes and disappear into a room together like it means something.

I hate telling my guy friends about any woman I meet because the first question is always, "Did you fuck her?" And followed by, "Why not?"

I don't know how to explain it except this—I'm just not wired that way.

I don't chase bodies. I don't hunt notches. I look for connection. For kindness. That's the sexiest thing in the world to me.

If it turns into something more, great. If it doesn't, that's great too.

What I am here for is simpler.

I'm here to be myself—the only version of me that I don't hate right now. To dance like an idiot. To sing too loud. To have stupid, joyful, unmitigated fun.

No shame. No filter. No care in the world. No expectations attached. No back pain. No stress. Nobody to impress. Nobody to offend. Nobody keeping score.

Not like back home.

And yeah—I splurged. I'd saved all damn year for this godforsaken trip, and I was going to spend it exactly how it felt right.

Only the real ones know this: ringing the bell is a universal signal—drinks for everyone in the bar. Instantly, the place erupts. Cheers. Laughter. Smiles from strangers who suddenly feel like family.

The bartenders, the waitresses, the random guy at the end of the bar—we're all long-lost brothers and sisters at an impromptu reunion.

Back home, I deploy that move sparingly. Drinks are damn near twenty bucks a pop now. The last time I rang a bell stateside, one round ran me five hundred bucks—tip not included.

But here? Drinks are two-fifty max. No need for tip.

I can ring that goddamn bell all night long and still walk out only a few hundred dollars lighter, depending on the bar.

Best investment in instant friendship.

I walked into the bar I saw earlier and immediately rang the bell. Happiness all around. It was a wild sight—apparently unbelievable. Old white guys sitting around either glaring, snapping photos, or taking videos of my dumbass in a tank top, dancing like a fool and singing songs I couldn't pronounce except for Sabai Sabai, which I played on repeat.

V was clearly taken with me.

She clung tighter than the rest, spoke better English too. Big, curious eyes—nothing like mine—always watching, always smiling. Shoulder-length hair, deep dimples, an easy warmth about her. She smelled as good as she looked. Perfect teeth, too—something you don't see often in Southeast Asia, where braces are a luxury most people can't afford. I know, I used to have really fucked up teeth. Probably why I'm obsessed with them.

I was, of course, buying endless shots. The only dude dead center on the dance floor, surrounded by V and a small swarm of girls—six or seven at one point. Even the bartender jumped in. They abandoned the old white guys without hesitation, which pissed off a few and thoroughly entertained the rest.

I was drenched head to toe in sweat. Didn't matter. It helped that Asian men don't exactly come with body hair or odor—otherwise I'd have cleared the place out like a durian bomb. The girls didn't care. The drinks were flowing, the money was moving, and the fun was thick in the air.

V held on even tighter through the sticky chaos, locking eyes with me the whole time. Smiling those deep, dangerous dimples I couldn't stop staring at. It appears, she couldn't either—she kept poking mine, laughing. A full-blown dimple contest in the middle of the madness.

I was grinning from ear to ear.

Then black.

I woke up the next morning in a panic—alone in my hotel bed, heart racing. Phone still there. Wallet still there. Even some cash, neatly folded beside me.

I had no idea how I got back.

Normally, this wouldn't scare me. It happens damn near every night. Somewhere along the way I developed a strange superpower: I black out, vanish, and still manage to materialize safely in my own bed. No witnesses. No memory. Just a hard cutoff point in the night where consciousness clocks out.

Friends say it's always the same. One minute I'm my usual jolly, talkative drunk. The next, I'm gone. A clean ninja disappearance. By morning, the evidence is predictable—Uber receipts, uber regrets, and an empty wallet.

But this time was different. This time I was in the middle of Bangkok. Would've been hard to get home even if I was sober. And my wallet still had money in it.

It could have been real bad…like bad, bad.

After my nood adventures during the day, I went back to the bar that night and asked Mamasan what went down. She said V took me home when I couldn't stand anymore. V confirmed it with a shy smile, so I handed her a couple thousand baht bills for her honesty. I was completely blown away.

She could have taken advantage of me. Taken everything I owned and left me for dead if she wanted. But she didn't even take a single coin. That to me was sexy as hell.

Then I asked if I could take her out for the night.

I paid Mamasan another thousand, grabbed V's soft little hand, and we took off into the bright, buzzing Bangkok night.

The next couple of days were dedicated solely to one sacred mission: eat at least one bowl of nood from every gigantic monstrosity of a mall in Bangkok. All of them—EmSphere, Icon Siam, Terminal 21, CentralWorld, Siam Paragon.

A full-on nood pilgrimage.

I didn't know it yet, but this was the first time in years my body and my soul were hungry for the same thing.

Chapter 14: Island Hopping

I met a fella at the corner Irish bar in Bangkok who said he was heading to Ao Nang.

I looked it up—perfect. Beaches, limestone cliffs, a good place to start before working my way through the islands and eventually up to Phuket, where I'd meet my Bestie in a couple weeks.

That night, I booked a one-way ticket and grabbed five nights at one of the few hotels still available.

The next morning, I was at Don Mueang Airport—the calmer, less chaotic, domestic cousin of Suvarnabhumi. I wish I'd learned that sooner.

I took a taxi three hours early for a side quest: finding Joy's noodle shop—the one I'd seen on Gary Butler's Roaming Chef YouTube channel. No address. Just clues. Across the street. On a corner. Mixed in with other shops. Eight corners total.

Odds were in my favor.

Sure enough, there she was.

Joy was the cutest auntie imaginable—humble, joyful, effortlessly warm. Exactly what I expected. And then some. She even mentioned she has a sister in Reno who owns a Thai restaurant. When I get back, I'm asking every Thai joint I love if they know her. Another side quest waiting to be unlocked.

I ordered two bowls—the best noods I've ever had in my life. Pork noodles and khao soi. $2.50 each. Unreal. I spiced them just right so they landed where it matters—straight in the soul.

It was so good I stopped mid-slurp and typed this into my phone at the plastic table, so I'd never forget:

Her name is Joy—and she brings me the most joy in life.
She owns the world's best noodle shop, tucked into a dirty alley under a rattling train station.
The only other woman I'll ever need besides my mom.
I think I'll move here now.

Looking back, that might've been the moment I doomed myself—chasing a bowl of noods better than hers. A mission. A dare. A lifelong bet.

I finally boarded the short flight to Krabi, spilled out with the tourists, and caught a taxi. A kid riding shotgun translated for his dad. Good kid. I tipped him a hundred baht for candy. He almost cried..

Forty-five minutes later, I checked into my hotel. Nothing fancy. A bed and a bum gun. No purple tea. But I was lucky to have a room at all.

I went hunting for noods but didn't love what I saw, so I settled for grilled tilapia. Solid. Noticed more Muslim Thais down here—makes sense the farther south you go. Malaysia was right next door.

That night, I wandered the main strip, landed at a live-band bar near my hotel, drank, listened, a little debauchery, and called it early.

The next morning, I woke to mangos slamming onto tin roofs, birds screaming like it was mating season, and motorbikes howling down the street. Could be worse, I thought. I could be freezing in Reno.

I needed noods—but tuk-tuks twice a day, just for noods, added up fast. Walking was tough with my busted back. Renting a motorbike terrified me. Ao Nang was busy. Traffic tight. Excuses stacked neatly.

Instead, I took a long-tail boat across the street to Railay Beach—only reachable by water thanks to towering limestone cliffs. I explored, swam, drank, ate a yummy bowl of beef noodle soup, and left before dusk to beat the rush.

Back in Ao Nang, I booked an island-hopping tour for the next morning, wandered the night market, hit the same bar, and stumbled home.

The tour hit all the classics—Monkey Beach, Maya Bay, Phi Phi, Bamboo Island. Beautiful, no doubt. But crowded. Too crowded.

That's when I realized something important:

I don't like busy places—no matter how pretty they are.

I met a cool blonde from Vancouver who kept me company all day. She'd survived a hardcore meditation retreat with monks in Chiang Mai and was heading to Phuket for EDC.

She asked my plans.

Same answer as always: none.

We exchanged numbers. Never met up again. She had friends and parties. I had drinks and solitude. Fair trade.

When I walked her to her scooter, I asked if her scooter game was strong enough to give me a ride back.

She laughed. "Not a chance."

I waved her off, hailed a tuk-tuk—and that's when the voice in my head, sounding exactly like Austin, said:

She's tiny. She's fearless. You're a little bitch.

I walked straight to the rental shop next to my hotel.

Rented the first bike they had. Bad breaks and all.

"No gas," they said. Shit.

Of course.

I had to ride through the busiest part of town just to find fuel. Terrified, sweating, stopping every few minutes to ask for directions. They kept pointing left. Always left.

I finally found the always-left gas station. Filled it up for a couple bucks.

And with frustration now stronger than fear, I blasted back through town—then kept riding. All over Ao Nang. And then again. Just to prove a point.

That night ended the usual way—my favorite bar, my favorite bartenders, dancing and singing with strangers.

And goddamn it…

I'm not a little bitch anymore.

Chapter 15: The Durian Farm

I woke up in Ao Nang with the meanest nood craving imaginable.

Head throbbing. Empty bottle of Mekhong on the nightstand. Sunlight crawling through a narrow slit in the blinds. Motorbikes howling like hungry dogs. The birds were angry at each other like bitter lovers. And those goddamn mangos.

Only one cure for a hangover like that: perfectly sweaty noods.

I checked Google Maps and found a noodle shop up north where the locals seemed to congregate. Hopped on my bike and tore down the quiet beachside road with salty air in my face and good fortune riding shotgun.

Too early. The noodle shop was still asleep.

But the caffeine gods were awake.

A tiny café stood nearby—a few small tables, a few chairs. Inside, an extremely tan Thai man sipped coffee and ate a slice of cake like it was his final meal. He was chatting with a petite Thai woman behind the counter—short black hair, skin painted white as rice, and a smile big enough to disarm an army.

He waved me in.

"Sit, Brother! Enjoy!"

He acted like he owned the place—grabbed the single-page menu and handed it to me. The woman smiled, bowed softly, and greeted me with the gentlest "Sawadee ka" I'd ever heard.

"Sawadee krup," I replied.

The man introduced himself as Yan. I told him my name and asked her's.

"Tik," she said—so softly I had to ask again. Always with that smile.

I ordered an Americano. Black coffee is a truth you can trust—especially when you're lactose intolerant and unsweetened by life.

Yan was friendly by profession—a taxi driver. Said he'd just driven a couple of Brits from Phuket all the way to Railay Beach. Five hours, he said. Traffic. I'd find out soon enough.

He told me Tik was his sister-in-law. Single. This was her shop.

I turned to Tik and said, "You're so pretty—your smile is more beautiful than all the temples in Chiang Mai! Why are you single? Bad breath?"

I made a huge mistake.

Her smile exploded—then vanished. Terror. Pure terror.

I forgot Thais don't do sarcasm.

I apologized a dozen times. Yan nearly choked on his cake laughing.

He asked Tik to bring me some cake too. It was sweet—but not as sweet as that small kindness. Sometimes I let myself take in moments like that, especially when accompanied by a soft and pretty face.

Yan told me his story.

Born in southern Krabi. Moved to Bangkok for school. Learned fast that walking beats traffic. Lived in a cheap apartment with thin walls and echoing hallways. His future wife lived there too—from Krabi as well. Every day he'd "just happen" to be outside when she passed. Asked her out relentlessly.

After a year of harassment, she finally gave up.

They married after graduation. She worked hotel reception. Yan tried law, failed upward into paralegal work. Traveled constantly. Gone more than home. Eventually she'd had enough of a husband who never showed up.

She made him quit.

Yan traded law books for the driver's seat of a taxi.

Bangkok broke them both, so fifteen years ago they escaped to Phuket—close enough to family, far enough to avoid entertaining every weekend. She became a hotel manager. He kept driving. No kids. No time.

He shrugged.

"It's okay, Brother. I can't get it up anymore anyway. Problem solved."

Phuket's crowded now too so he's counting down to retirement on his durian farm back in Krabi.

Three hundred trees. Five years old. Ten fruits per tree right now.

But at ten years?

A hundred fruits per tree.

He did the math like a man who's rehearsed it alone for years:

$10 per fruit × 100 fruits × 300 trees.

Three hundred thousand dollars. A big number anywhere in the world.

Convert it to baht and his grin widened. Ten million-plus a year.

Then he lit up even more. He planned to open an Instagram-famous café at the farm entrance. Tourists. Lattes. Durian selfies.

Only problem?

No English-speaking manager to charm the tourists.

"You're speaking to him, Brother!" I yelled.

His face exploded with joy. We hugged like kids making an impossible promise.

Then I hesitated.

"Wait… Brother… I only know how to make black coffee."

Yan smirked.

"That's okay," he said. "You marry my sister-in-law."

The bastard.

Tik blushed. I said deal. We shook on it.

Another path unlocked. Beckoning.

The noodle shop never opened. Wrong day, I guess. Google Maps lies out here. So I ate chicken porridge instead—minced chicken, over-easy egg, a buck fifty, plastic table, plastic chairs, dirty road.

I waved goodbye to Yan and Tik and headed out.

A tour shop run by an old man in a turban pointed me toward the Emerald Pool. Called his little brother. Little Brother drove me. No English. We Google-translated the whole way.

The pool was… emerald. Cool. But what I loved was the drive. Where Tik and Yan grew up.

Jungles. Dirt roads. Coconut trees. Rubber plantations. Impossible limestone mountains like watercolor paintings. Buffaloes. Shacks in rice fields with banana-leaf roofs.

It made me happy.

Then it hit me:

I was born in a place like this.

A hut. Jungle. Mountains. Dirt floor. Banana-leaf roof.

That was the backyard of my happiness.

Maybe I'm supposed to come back to it.

On the way back I asked for the best boat noodles in Krabi. Two bowls. Sen lek and woon sen. Skinny and clear noods. Ate silently with Little Brother, smiling.

Mission accomplished.

That night I drank Changs on the beach by myself. Reflecting.

The next day was slow. Porridge. Tik's café. Black coffee. Smiles. Awkward stares. Her English wasn't that good and my Thai was even worse. It's fine because I'm quite comfortable just quietly staring and smiling.

I asked her to run away with me.

"Mai dai.," she said, cannot, pointing at her new shop. Divorced too. Not about to make another mistake with a drunk halfway across the world.

I asked for her WhatsApp. Asked her to be my girlfriend.

She said, chai. Yes.

Promise? Chai.

That afternoon I swam. Ate too much. Thought:

What a great problem to have.

Then:

How do I make this my problem every day?

By nightfall, I knew Ao Nang had given me everything it had. I booked the next island—Koh Yao Yai.

A good place to disappear.

Chapter 16: Koh Yao Yai

The next morning, I went for one last coffee with Tik. She let me sit there and stare at her for a whole quiet hour—between the few customers who wandered in—while I slowly killed two Americanos and picked at cake like I had nowhere to be. An hour of her life she'd never get back.

I don't think she minded giving it to me.

When I told her I was leaving, her face did that rare thing good people do—sad for themselves, happy for you, both at the same time. No selfishness. No guilt. Just a soft kind of acceptance.

And her usual answer to everything, "Mai pen rai". No worries.

I asked her one more time to run away with me. Again, she said, "Mai dai."

Fine. I tried.

Before I left, I made her promise not to marry another guy—that she was all mine. She laughed and said, chai. I hugged her, kissed her cheek, then asked for one back. She pecked me on the lips. A little smudge of lipstick stayed on me.

I carried it longer than I should've.

After that, it was go-go-go. Returned the bike. Grabbed my bags. Tuk-tuk to the pier. Then the speedboat—hot, cramped, the air thick with bodies, salt, and humidity. I sweated like a soggy sock until the boat finally tore through the water and the wind punched clean through the cabin.

Forty-five minutes of pure, salty freedom.

Koh Yao Yai grew larger in the distance. Peace. Quiet. A long exhale I'd been holding.

We landed on a long, empty pier with a line of motorbikes parked in the sun—workers' bikes, not tourist toys. Two kids were fishing at the edge: sandals, sun-browned legs, huge smiles. No fancy rods—just sticks and lines, a small bucket already half-full of tiny silver fish flashing like coins.

The sea was so abundant it felt like a cheat code. The moment they dipped bait in, a baby fish bit down and they yanked it up with bare hands, laughing like the ocean was their personal candy jar.

Simpler days. Happier days. Days when life didn't need much to feel whole.

I asked if I could take a picture with them. Their faces lit up. One of them held up a fish dangling from the line, joyfully smiling like I was the celebrity.

Then I shared a tuk-tuk with a European mother and daughter from the same boat, headed the same direction.

On both sides, rubber plantations stretched out like endless cathedrals of green—rows of trees standing still and solemn, their trunks cut with old diagonal scars that glistened with sap. Little wounds that fed entire families. Sunlight slipped through the leaves in thin golden blades, flickering across my face.

Even the air felt different here—cleaner, thicker, sweeter, quieter. Like it knew how to hold stories.

Every few minutes, a mini village appeared: wooden houses on stilts, a blue tarp, a dog too lazy to bark, wild chickens minding their own business. An old man rocking in a hammock. Barefoot kids chasing a rubber ball. A woman chopping coconut with the calm precision of a magician. A man sharpening a machete on stone like it was meditation.

The island wasn't in a rush.

For once, neither was I.

As we moved farther south, the hills rose—gentle at first, then sharper—lifting me into views that felt almost unreal. The ocean appeared and disappeared through breaks in the trees. I'd blink and see deep blue, then blink again and it was swallowed by jungle.

Every turn gave me something new: a sudden stretch of coastline, a fence of coconut trees bowing toward the sand, a water buffalo standing knee-deep in mud, staring at me like I was the one out of place.

We dropped the ladies near the big town at a retreat place, then kept going to my last-minute hotel—the one I'd booked half-drunk the night before.

It was almost too nice. Big. Polished. Impressive in all the wrong ways. Cheap by American standards, but huge and crowded—one of those places with a giant maze layout and an enormous swimming pool that slithers past everyone's patios. Kids splashing everywhere. It tugged at memories—good ones, complicated ones.

Not my kind of place.

But it was the cheapest thing near the beach I could find.

And they had purple tea.

The next morning I went hunting for a bike rental. Every shop near the hotel was empty, or the broken bikes had nobody around to fix them. Nothing left.

That's how I stumbled into Lok—posted up out front like he'd been waiting.

He tried to sell me tours. I told him I'd already done most of them. I said I only needed one thing:

A motorbike.

He shook his head. "Very busy, brother. Sorry."

I must've looked genuinely crushed because he waved me back.

"No problem, brother," he said. "I give you mine."

He said it was old, so he'd give me a deal.

I didn't care about the deal.

I cared about wheels.

It wasn't comfortable. It wasn't shiny. It rattled like a tin can full of loose screws.

It was perfect.

There is nothing better than riding an old motorbike around a remote Thai island all day—stopping at tiny cafés, hole-in-the-wall restaurants, mom-and-pop shops, random bars, waterfalls, deserted beaches. Talking to strangers. Eating whatever they hand you. Drinking whatever they pour you.

In a hurry to get nowhere.

Name a better way to live.

I won't wait—I've got more exploring to do.

And screw those big, shiny, comfy bikes the farangs rent. Give me the old ones that locals have been riding for decades so I can feel every bump in the road. I like my ass sore after a long ride.

Earned pain. Happy pain.

Only now am I starting to understand why they call Thailand the Land of Smiles.

I rode from the very bottom of the island to the very top, and every man, woman, and child smiled at me as I passed—no matter what they were doing: walking the road, riding their own bikes, cutting rubber, tending buffalo, cooking food, mixing sodas, sitting outside their huts gossiping with friends.

They'd look up and smile. Or press their hands together and whisper, "Sawadee krup," if their hands weren't busy.

It's warmth—but it's also curiosity. It took me back to the refugee camp, the first time we saw white people: French nurses doing UN work at the makeshift hospital. We stared the same way—wonder braided with caution, awe tangled up with confusion.

Now I'm the farang.

Oh, how things change—and somehow stay the same.

I always prefer old local shops over fancy English menus and Google reviews. If I see a row of motorbikes parked outside a shack with not a single white person in sight, that's my cue:

This is going to be incredible.

My pointing and hand gestures got laughs—kind laughs—but I didn't mind. I'm learning that laughter is its own language.

By late afternoon I rode down to the beaches and stretched out in the shade and started writing—meaning pounding on my Notes app like it owed me money.

This is where I can think.

And because I can finally think again, I can write again.

I haven't written in a decade. Turns out I only write when I'm inspired, and I'm only inspired when I'm on an island. Back home? Nothing comes out. Just blank.

So fuck it. I'll travel more. I'll write when inspiration hits. I'll jot down whatever I can—whenever I can—about whoever and whatever moves me.

Maybe this is the only way I know how to write.

Maybe that's okay.

For the next couple days, I rode and wrote.

And the island kept whispering, quiet as a warm tide:

Slow down.

You're safe here.

Keep going.

Don

His name was Don. He waved me down as I rattled past on my creaky motorbike. A Muslim Thai man with a grin like sunrise and one eye pointing slightly left. He ran a coffee-and-smoothie shack on the side of the road—simple, but somehow elegant. Hand-carved wooden furniture. A roof that listened when it rained. He built it himself after his father died and left him a small slice of land—and all the ghosts that came with it.

His family still bleeds rubber for a living. Thousands of trees. Sap washed and folded day after day like time itself.

But Don wanted something smaller. Something quieter.

Three generations of rubber was enough. So he traded endless fields for a few square meters of peace. Opens at ten. Closes at four. Decides the day is enough when the sun says it is.

His wife doesn't speak a word of English. She moves like a quiet shadow between the counter and the blender, doing all the work. Don talks to strangers all day and laughs with them. Says it's better than talking to his wife.

That's probably the secret to a long marriage—silent, steady, and true.

She made me an Americano that tasted like hope and a coconut smoothie so naturally sweet it made me emotional for no good reason.

Don thought I was Korean—said my eyes gave it away—but the sun had darkened me too much, so he decided I must be mixed.

"Like everything good," he said.

He switched to broken French for a couple from Marseille, tripping over oui oui and merci like a kid showing off.

And I thought—maybe this is it.

Maybe the dream isn't money or fame or the next big goddamn thing. Maybe it's a small coffee shack in paradise, a woman who knows your heart without understanding your words, and a life measured in smiles from strangers, one cup at a time.

Thanks for the blueprint, buddy.

I might've met the girl a couple days ago.

The Field

The volleyball isn't great. The net isn't even that high. But I love how they play like it's the World Cup.

Full uniforms. Full effort. Loud commentators taking their jobs way too seriously in the best possible way.

And the whole town comes out to watch. Kids. Elders. Shopkeepers. Moms with babies. Guys still in work clothes. Everyone circling a dusty court like it's the beating heart of the island.

It takes me straight back to the refugee camp.

My dad and uncles played soccer every afternoon on a dirt field in the middle of the camp in the middle of nowhere. I'd stand on the east side—too lazy to walk around to the shade—squinting into the sun, watching my dad run.

I swear the wrinkles on my forehead were earned right there.

Back then, there were no farangs.

Now I'm the only one—but with my deep tan, nobody noticed.

The laughter. The dust. The way they cheered for players they know by name.

Funny how a court halfway across the world can peel your whole past open.

The Perfect Day

I said earlier there's nothing better than riding a motorbike all day on a remote Thai island.

I take it back.

Nothing—and I mean nothing—beats a perfect beach day.

Seventy-five degrees. Humidity barely a whisper. Shade from a mangrove tree shaped like God designed it personally. A breeze that knew when to show up.

The sand was perfect—not too soft, not too sharp. I walked where the tide kissed the shore: damp, cool, firm. Even the pint-sized sand crabs with smiles on their bellies agreed. That's where they built their miniature fortresses, popping out like they owned the place then dashing back inside when I stumbled by.

The water was a degree cooler than the air. Clear as glass. Fish everywhere—tiny sparks, long shadows, flashes of color. Chest-deep was the sweet spot. All of us floating together, weightless.

This water is medicine.

The back pain I carry every day disappears the moment I float. Gone. For people who live with pain, that relief becomes everything. You'd walk through hell for it.

I just walk fifteen feet into the sea.

No noise. No Russians. No influencer photo shoots. No crying babies. Just waves, leaves, cicadas.

A Mai Tai. A watermelon smoothie. A small water.

The holy trinity.

I got a massage—picked right: young, strong, soft hands, no sharp elbows and tickly fingers. I snored.

That's how you know it's good.

I tipped her. Not because you're supposed to—but because gratitude matters. Her smile was worth it.

Two thoughts drifted through my blissed-out brain:

Maybe I should propose to her.

Then go find noods.

It was already two p.m.

Where the did the day go?

Lok

Sunday's his only day off. He spends it working on his house by the pier—shirtless, sanding, staining, smiling quietly. The air smells like salt and sun-dried fish.

Lok's a tour promoter, part-time tuk-tuk driver, full-time good guy.

Speaks better English than he admits.

He rented me a scooter for two days—400 baht. Filled it up. Eleven-fifty for freedom. Show me another country where that's possible.

He loves fishing and painting.

So do I—though I'm useless at both but I make a lot of money painting.

He said, Mai khao jai—he didn't understand.

I said neither did I.

We laughed.

That laugh turned heavy.

Same heartbreak. Different details.

His girl left for someone "better" than a tuk-tuk driver. He learned English. Made more money. But said he can't get it up anymore, so he stopped looking for love.

That might've been the saddest poem I heard all week.

I said, Mai pen rai. No worries.

And for the first time in years, I meant it. Not intellectually. Not as a mantra. In my bones.

It had been almost two years since she left, and I don't think back with sadness or rage anymore. Not the rehearsed bitterness. Not the gut-tightening loops of could'ves, should'ves, why-the-fucks.

Most days, I don't think about her at all.

And when I do, it isn't grief or fury—it's something that feels suspiciously like understanding. Maybe even forgiveness.

Which is strange. Strange because I never thought I'd get here. Strange because time didn't fix it—*I did*, by walking away from everything and going hunting for whoever I am now.

And strangest of all: for the first time since she left, the story isn't about her anymore.

It's about me.

I was holding the keys to this prison of tears the whole time.

I rode every inch of that island. Made drunk promises to it I can't remember but still carry somewhere deep.

And every time I think of Koh Yao Yai, it whispers back:

Mai pen rai.

This was the most peace I've felt since that hammock on Koh Phangan. If this is all life ever gives me—these brief windows of stillness—I'll take them.

I'll just keep coming back. No matter the cost.

Something inside me finally unclenched. The noise dropped away.

I don't know exactly what I'm chasing.

But I'm close. Close enough to feel it before I see it.

One island at a time.

One mountain a day.

And when I finally find it—I'll know.

Chapter 17: Phuket

I woke up, grabbed the shuttle to the pier, and hopped a speedboat to Phuket.

Yan was already waiting—big goofy smile, arms wide—scooping me up like family. We grabbed lunch, traded more stories, then he drove me to his house on the east side of town. Cleaner and nicer than I expected. He paid three million baht for it—about eighty-five grand—and walked me through every room like it was the White House.

Then he brought me to his favorite restaurant next door, introducing me to everyone like I mattered.

That man's heart is made of gold and coconut husk.

We ate heavy. Yan sipped one beer slow. I crushed Changs and chased them with Mekhong like I was still bulletproof. Then my phone chirped.

Bestie lands in an hour.

Yan had already paid the tab and wouldn't let me touch it. I bowed to everyone like I was leaving a royal court and we headed for the airport. He dropped me at international arrivals and waited in short-term parking like a damn guardian angel.

I'd forgotten how tall she was. In Phuket she looked even taller—like the humidity stretched her out another foot. Her smile is usually guarded, earned, but today it was wide open. I was happy for her. And selfishly happy for myself, because for once I wasn't traveling alone with only my intrusive thoughts for company.

I called Yan. He swooped in immediately.

Group Therapy had planned some glamorous resort getaway, but everyone bailed except Sweet Tea, forcing her to rebuild the whole thing from scraps. Resilience is her personality.

The drive should've taken thirty minutes.

It took well over an hour and a half.

EDC traffic turned Phuket into a parking lot. Yan nearly got us flattened more times than I could count on the way to Patong. Tea clutched door handles and headrests like she was praying to every god at once.

I trusted Yan.

He needs me alive for the durian empire.

When we finally arrived, I hugged him like he'd delivered me from war.

The "hotel" was basically a motel.

Next to a cemetery.

Cheap. Loud. Wrong in all the right ways. Perfect for me. Only problem: one king bed in the center of the room. Fine for one night. Horrific long-term. I slipped the receptionist 500 baht and asked about upgrades.

He grinned and told me the truth—the king was two full beds pushed together. He'd tell the maids to split them in the morning.

Deal done.

I woke early. Tea was still asleep on her side. I slipped out, grabbed coffee, a Red Bull, and—because balance—a bottle of Jameson. Nothing else was open, so I sat beside the cemetery sipping my drinks, wondering if my grandpa would love or hate this place.

Knowing him, he'd hate it as much as I did.

Mosquitoes swarmed me like old lovers.

"Hold up," I told them. "My blood's old, crusty, and loaded with alcohol. Go try Sweet Tea."

Didn't work. They bit me anyway.

When Tea finally woke, we did her ritual.

She always takes a tour first.

Our first was a Phuket food tour—pure sensory overload—but the guide stole the show: a thick Thai accent tangled with a full-blown British one. It broke my brain. I laughed the whole time.

Later I asked how she got that accent.

"That's how I was taught," she said.

UK tutor.

Mystery solved.

People parted as we walked. Tea nearly taller than all the temples. Me beside her like some happy mascot.

You couldn't make this shit up.

Then came the elephant tour—non-riding. I never wanted to go, but I did. I've always been uneasy around captive elephants, even the "ethical" kind. Their eyes hold too much memory. Even the babies carry sorrow. Gentle giants. Peaceful. Powerful. Ancient.

Then the ping-pong show.

I'm not describing it.

Children don't belong there.

Fuck you, Russians.

Traffic made everything worse. Tuk-tuks were expensive and slow. Everywhere was gridlock.

Phuket sucks ass.

I drank more than usual. The days blurred—Irish bars, biker bars, beaches, noise. Old white guys stared at me like:

How the hell did this short jackass land that tall woman?

I shrugged back.

Hell if I know, brother.

We avoided shellfish because she's allergic—though I still can't remember which ones she can or can't eat. We kept it easy.

Then reality showed up.

My assistant called complaining it took three guys a week to paint twenty vases—a four-hour job for one man.

My vacation snapped in half.

I'd planned Chiang Mai next.

Not happening.

I hung up and said, "Welp. Guess I'm going home too."

Tea was leaving the next day anyway.

Responsibility waits for no one.

After checkout, we shared a taxi to the airport. I dropped her at international departures and headed to domestic—back to Bangkok.

One last night.

One last bowl of noods.

I surprised V and the girls.

Drinks for everyone.

We sang Sabai Sabai until morning.

Then it was back to hell on earth.

Chapter 18: Hell On Earth

Same damn thing every time I come back home.

Jet lag.
Sick.
A body that doesn't know where it is.
And nobody cares.

If anything, people look irritated that you're back at all—like they hate you for leaving, and hate you more for returning. Maybe it's the contrast. The cruel, stupid, violent contrast. One month in warm golden paradise, where every moment feels like possibility, then you're dropped face-first into cold, gray, breathless impossibility.

One minute you're on a beach the color of honey, warm water wrapping your ankles like silk, air brushing your skin like it loves you. The next minute you step out of the airport into a cold, dry slap from the universe.

Your body goes into shock. You get sick the moment the cold touches you—like your soul rejects the place on contact. Coughing. Aching. Shivering.

And your back pain returns like an old enemy kicking down the door.

Miss me, motherfucker?

It grabs you by the spine and shakes the life out of you until you can't breathe right, can't think right, can't stand right.

Then—insult to injury—you still have to drag your sorry ass to work. You still have to stare at miserable faces you can't stand. People bloated with bitterness. Everyone suspicious. Everyone talking shit. Everyone pissed you dared to live better than them for even a moment.

And the bullshit starts immediately. Same lies. Same laziness. Same egos. Same drama. The same bastards doing the same bastard things.

The same shitshow I left behind—waiting for me like a jealous ex who never moved on.

The moment I got back, all hell broke loose. The owner made a string of stupid decisions in my absence, then used the money he owed me to build himself a fancy new office.

Perfect.

And then I learned my friend—someone I trusted, loved, someone I hired when he was down—stabbed me the first chance he got.

Also perfect.

Just business, right?

Fuck you too.
Thanks for finally revealing yourself.

My back got worse by the day. Every hour heavier. Every morning stiffer. Every night ending in ice packs, heating pads, and groans swallowed into the dark. Running that place felt like herding blind cats with egos the size of GMC trucks. One fire here, three fires there. I had to cut loose a few just to keep the ship from sinking.

And the more crap I dealt with, the worse my back got—like it was allergic to this place.

That's when I finally admitted what I'd been denying.

I needed surgery.

I should've done it the year before, but everyone told me the same thing: "You're still young. Hold off."

As if pain respects youth. As if suffering has an age requirement.

They hired a new Operations Officer, and of course—the company tried to screw him too. Patterns don't change. People don't learn.

And that's when the truth started leaking out of the cracks: unpaid wages, criminal shit, shady books, dark corners with darker intentions—lies stacked on lies. I watched them brag about not paying the new officer what he was owed, and something in me finally clicked:

If they'll do it to him in front of you, they're doing it to you too.

I worked through all of it, but the pain kept climbing. Walking became a negotiation. Breathing felt like a contract I never signed.

The stress stacked like bricks on my spine. The drinking got worse. The Advils stopped working, so I started swallowing more. Five at a time. Three times a day. A bottle a week. My stomach hated me.

My back hated me more.

Then the numbness started.

Left leg—gone.
Dead.
A useless hunk of meat dragging behind me.

For a month I couldn't feel a damn thing below the knee.

Apparently, I hid it well. Too well.

A few people noticed—the rare ones who caught me drunk enough, cracked open enough—but most had no clue. On the outside, I looked fine. Happy, even. Posting ridiculous smiles on Facebook like I was living my best life.

Inside?

I was absolutely miserable.

Spinning in a revolving hell with no exit. Working with some of the worst people ever assembled under one roof. Pretending everything was fine while breaking one quiet inch at a time.

I waited nearly six months for surgery. Six months of limbo. Six months of insurance bullshit—delays, approvals, rejections, expirations. Relief kept getting shoved further away like a cruel mirage.

That's all I was waiting for: one clean breath without pain. One moment where my body didn't betray me.

While the calendar crawled, I stayed trapped in a place I hated, surrounded by hollow souls whose voices hurt worse than the injury itself. Every morning, I walked into that building like a condemned man clocking in for his final shift.

Every day was the same:

Endure.
Endure.
Endure.

One mountain a day, Cali.

I wasn't living.

I was waiting.

Waiting for someone to cut into me and free me from the demon devouring my spine. Waiting to remember what it felt like to be human again.

And that kind of waiting changes a man. Quietly. Slowly. Until something inside whispers:

You're not meant to live like this.

My body didn't whisper.

It screamed.

Walking from the car felt impossible. I moved like a hunched corpse dragging a nearly severed limb behind him.

Then the morning finally came. Mom and Dad showed up the day before surgery. My apartment wasn't fit for recovery, so they booked a hotel nearby. They sat with me like sentries, holding the line.

The nurses prepped me. Stripped me down to a thin gown that offered no dignity and even less warmth. The anesthesiologist asked questions I barely answered.

Then the surgeon arrived—calm, steady.

"You're gonna be fine, Kao."

And then—

lights out.

Chapter 19: Red Hot Knives

When I opened my eyes, the first thing I noticed was the absence of pain in my left leg.

Not "less pain."
Not "different pain."
No pain.

A clean shock of joy hit me so fast I almost didn't trust it. I wiggled my toes. Moved my foot. Again. And again—just to be sure. My lower back was still sore—tender, swollen, pissed off like it had survived a war.

But the old pain was gone. Not the sharp electric bolt that stole my breath. Not the burning, gnawing line of fire ants crawling down my leg.

Just soreness.
Just healing.
Just silence where agony used to squat.

The relief I'd been praying for—begging for in the dead hours—arrived quietly, like a long-lost friend slipping back into the room and sitting beside me without a word.

Later the surgeon came in, shaking his head. "Relax," he said. "You'll be fine. But it was bad. Really bad. Usually takes me an hour—this one took three. One of the worst backs I've ever seen on someone your age."

"Thanks, Doc," I said. "That makes me feel real young and spry."

He laughed. Told me my mom was a sweetheart. Told me I was healthy. Told me I'd get through it. Then he gave me his personal number—said call if anything went sideways.

That kind of care stays with you.

He asked if I wanted physical therapy—pay someone to babysit me on a treadmill every day. Hell no. A treadmill is a conveyor belt of misery. I'd walk myself—on a farm, on an island—anywhere but that torture device.

It was 12:30 a.m. Wednesday.

Day 1 of recovery.

The fentanyl was long gone. The morphine too. Two hours until the next shot, and I was sprawled in bed like I was rolling across a field of red-hot knives. No position worked. The gown scratched my skin. The sheets felt like sandpaper.

If I got one single hour of sleep, the angels must've clocked in and taken overtime.

So I wrote. In pain. Then edited in relief. Documenting the next chapter of my life in real time:

For over two years I'd crawled through a cold trench of quiet defeat—a place no one talks about, where hope becomes a rumor you stop believing in.

But now something flickered.

Small.
Fragile.
Real.

My mind—once a mangled mess of broken glass and barbed wire—started stitching itself back together. Slowly. Painfully. Miraculously. Bestie didn't believe me when I said the psychological lesions finally scabbed over. The gaping wounds became scars.

Beautiful scars.
Proof I survived again.

Soon enough, the wreckage of my body would follow. And I'd rise—like a goddamn phoenix—rebuilt, forged in pain and sorrow.

The world better be ready.

The next morning, the nurses—my guardian angels—came to release me and laid out the rules of my new life:

No BLTs.
No bending.
No lifting.
No turning.
For six months.

They made me prove it by putting on my socks—lifting the foot up to meet me and not bending down to meet the foot like before. It was ugly—comical even—but I managed. Barely. They stood there patient and proud like I'd just climbed Everest, then gave me a high five.

And it hit me: I had to relearn everything. How to get out of bed. How to go potty. How to tie shoes. How to walk. How to function. How to be.

And instead of bitterness, I felt something else.

Relief.

Because this wasn't punishment. This was…

A reset.
A new beginning.
A new man with a new lease on life.

The moment the pain vanished, the truth came in clean as a whistle:

When darkness lifts, you finally see everything.

Dad asked if I could make it to Willows. "If you can do it, it's better than here," he said. "We'll take care of you at home." He had a point. We stopped at my place, packed for a week, and drove. Every mile was torture. Every bump a threat. Every bathroom stop necessary—the nerves needed time.

We made it by dusk. Mom cooked the best fresh Hmong chicken I've ever tasted in my life.

That night I tried to sleep.

Failed.

I spun around for six hours, half-delirious, and at some point I grabbed my phone and wrote:

Writing on the Wall

I read it once in a dim, shit-infested bathroom stall, scribbled on the wall like a prophecy whispered into my wasted youth:

When you're healthy, you've got a million problems.
When you're not, you've only got the one.

Here I am again, past midnight, rolling on a bed of red-hot knives. My buddy Jameson abandoned ship five days ago. Only Tylenol and scraps of willpower remain.

Pain makes decisions sharp and stupid.

But this didn't feel stupid. The clarity felt too real.

My parents would be devastated. They always are. They never understand that what they want for their kids isn't always what the kids want in the end. I could already hear my brother telling my nieces: "Your uncle is such a fucking pain."

I'm blessed—I truly am—to have burned through every dream I ever chased. Some beautiful, some grotesque, some doomed from the start. But I lived them all.

And in the end, a man doesn't need much.

A bed.
And a bowl of noods.

I have enough stacked to outlast this life. So I'll be fine.

I knew I'd made up my mind. Only two fears remained:

1. That I wouldn't do right by my boys.
2. That I'd bump into her again in the next life.

Last year I chose freedom. This year I'm adding health.

Freedom + health is a nuclear payload few can withstand.

Add the fact that I'm done defending, done denying, done arguing… And you've got a dangerous man about to do a dangerous thing.

Which I knew was quitting my stupid ass job as soon as I can and moving to Thailand—or something very similar to that.

But first things first.

Second things never.

Before anything else, I had to walk. You can't go anywhere if you can't walk. No airports. No planes. No Thailand.

Recovery had to start at home, with one brutal, humbling goal:

Make it around the goddamn block.

If I can make it around the block, I can survive an airport. If I can survive an airport, I can survive two or three. If I can do that, I can eventually get to Thailand.

I knew the danger. Push too hard and I'd reinjure myself—maybe worse than before. But if I moved carefully, deliberately—nudging past the edge—recovery might come faster.

Controlled stupidity.
Strategic recklessness.

I just had to reach the corner.

On day one, after breakfast, I laced up my best gym shoes like they were armor and opened the door.

Crossing the threshold felt ceremonial.

I limped down the driveway. Onto the sidewalk. Deep breath. Halfway down the first side, the pain lit up my lower back like fireworks. My left foot dragged behind me, stubborn and stupid. The numbness was gone, but the coordination wasn't. I had to command it:

Move.
Lift.
Follow through.

Like teaching a newborn limb how to exist.

That's the cruel part about nerve damage.

The pain isn't the worst of it.
It's having to tell your body how to do things it used to do without asking.

Step by step, breath by breath, I stood there sweating, shaking, bargaining with my own spine and thought:

This is it.
This is how everything starts again.
Everything starts with nerves.
And I will sharpen these nerves into steel.

Nerves shift. Nerves migrate. They jump ship when something gets pinched, spread the suffering around, then retreat when both sides are beaten raw. They're stubborn sons of bitches—hard to kill, but not impossible.

And when they die, they take almost everything with them.

One day you're doing backflips, running up stairs, walking a block without thinking. The next you're scanning for elevators like they're holy ground, and one block feels like a legislative hearing—slow, painful, negotiated.

Bones heal. Muscle heals. Quickly.

Nerves take months.
Years.
Sometimes never.

Lose your nerves and you lose everything. Without nerves, you're just meat and bone waiting for instructions that never come.

Nerves are what let you feel—love, pain, fear, courage. They're what let you take chances. They're what let you live.

And nerves give you the strength to do the impossible.

Like reaching the first corner on day two—my little nephew walking beside me, offering his shoulder like a handrail, pretending he wasn't there to make sure I didn't fall. Like reaching that same corner on day three—alone. Like circling the block. Like pushing farther to Uncle Roy's house on the third corner, just because I could and I had to check on the old man.

Because the impossible never shows up all at once.

It starts as a corner.
Then a block.
Then a life.

And one day, without asking permission—

It becomes a plane ticket.

I joked a little too much that week about moving to Thailand with my girlfriend to heal.

Joked—
but meant it.

Enough to scare my parents.

Chapter 20: The Promise

On the fifth day of my recovery, I woke up to the smell of something delicious in the air.

The best way to wake up.

It was like a full Cinderella reel playing in my head and soul. I dragged myself out of bed, took a piss, washed my hands, washed my face, brushed my teeth, and shuffled into the kitchen. Mom was already there, ladling khao poon into a big bowl like she always does when she wants to fix everything at once.

One of my favorite noods. Always the best when she makes it.

I sat down, sweaty, stiff, grateful. She sat next to me, wiped the sweat off my face with that familiar mix of tenderness and authority, and then—without warning—she pulled the pin.

"If you have ever loved me," she said, "you won't go to Thailand!"

"Mom," I said quickly, "don't you say it."

"I'll kill myself!"

Too late.

She went full theatrical. Full tears. No restraint. I've seen her cry a thousand times in my life, but this one? Award-winning. A masterclass. Decades of practice, perfected.

"Mom, I can't believe you just tried to pull that on me," I said, half laughing, half panicking. "It's not that serious. Just because it's all I talk about? Just because I write about it like a love-sick teenager on Facebook? Just because all I do is research how to live there? It's really not that serious."

She stared at me hard.

I folded.

"Fine. I won't go this year. Okay? Relax. Love you."

I kissed her on the head and escaped to the shower like a coward who knows when he's beaten.

I thought the show was over.

It wasn't.

That night, right as I was about to crawl into bed, Dad sat me down on the couch like I was twelve years old again.

"It's not a good idea you go to Thailand, Son."

"What the hell is this?" I said.

"I don't like it. And you're not gonna like it if you go to Thailand."

"Hold on," I said. "You're gonna disown me? I'm forty-six, Dad. Are you kidding me?"

But his tone told me everything. The serious one. The one he saves for death or when I've seriously fucked up.

No point arguing.

"Fine," I said. "I promise I won't go to Thailand this year. You happy? Love you, Dad. Goodnight. We gotta leave early tomorrow and beat the weather."

I hugged him.

And later that night, I sat alone and wrote.

My parents hated that place when they finally went back after 35 years.

I absolutely fucking loved it, to their heartbreak and tears.

They couldn't understand why the heat and decay,

the chaos, the corruption, the mosquitos' ballet,

the smell of diesel and sweat in the air

made me grin like a drunk lunatic without a care,

while it made them frown like the past had returned—

ghosts at the table, hard-earned lessons unlearned.

They spent the best years of their lives escaping that place,

to find peace, to rebuild, to erase every trace.

They wanted calm, they earned their rest,

and I can't blame them—it made them their best.

The thought of me going back, their once-prized son,

must burn like a cigarette that's never done.

Their peace comes, from distance from it, a world apart,

and never returning to the start.

I wanted to go back to what made me feel alive,

when love didn't need money to survive.

When being a kid and being held was enough,

before the world told me I needed more stuff.

Before America taught me "poor" was a curse,

and the more I had, the more I'd have to rehearse.

Before I forgot what joy could mean,

before I traded simple for glossy sheens.

My peace comes, from proximity

to it, always returning to the humidity.

Now that I've climbed each ladder they set,

touched every dream they helped me get,

I just want to sit under that same damned sun,

the one that burned us all when life had just begun.

Same kid who pressed his cheeks against those barbed-wire dreams,

now wants to walk those streets where the sunlight gleams.

When I told them that, I broke their heart in two

never seen them so angry, so lost, so blue.

Nearly disowned at forty-six years old—

Jesus Christ, this story never gets old.

They made me promise not to go back this year,

so I nodded, smiled, and begrudgingly said, "I hear."

I promised both I wouldn't go to Thailand.

But I knew something they didn't know.
There are plenty of other goddamn places in Southeast Asia with beaches.

Dad drove me home. We talked about everything and nothing.
He dropped me off. I hugged and kissed him. Then he headed back.

Healing was my only mission. I had to quit the hellish job.
I had to travel. But I couldn't survive a 20-hour flight yet.
I needed to practice.

So I tried somewhere close.
Somewhere simple.
Somewhere with noods.

Chapter 21: Minneapolis

Labor Day Weekend was supposed to be quiet. The plan was simple: horizontal therapy, day drinking, naps. One errand a day. Max. A slow lap around the block to satisfy physical therapy. Treat the new back like fine china—fragile, expensive, no returns.

I figured I'd hit the tired Rib Cook-Off for a protein fix, avoid the desert sun, and call it a weekend. But standing there with my keys in hand, something cracked. The ribs lost all appeal. The town felt wilted—sad, stale, gray—like it had been hungover from a life it never asked for.

All I wanted was noods.

And this town didn't have a single bowl worth craving.

So fuck it.

If the noods won't come to me, I'll go to them.

I grabbed my silver four-wheeled luggage—always packed, always ready—passport tucked into the side pocket, cash folded like a dirty secret.

Not prepared for disaster.
Prepared for escape.

I called an Uber to the airport—thank God it's a five-minute drive—walked to the counter and bought the first ticket to somewhere I'd never been.

Minneapolis.
A strange choice.
A satisfying one.

Three and a half grueling hours later. The Twin Cities—stitched together by lakes and winter—are where most of my people landed after fleeing jungles, mountains, and refugee camps. Jungle Asians in a frozen world, thriving where we shouldn't. We're like cockroaches you can't kill. Give us dirt and weather and we'll build a garden.

The secret?
We're the world's greatest farmers.

We can make anything grow. Anywhere. Except me. I can't even keep cilantro alive.

There's a Hmong saying: Anywhere there are clouds, that's where we build our homes.

And we brought the noods with us. Every kind. Pho. Khao piak. Khao poon. Bean thread. Vermicelli.

If it stretches, slurps, or swims in broth—we've perfected it.

My theory—baptized by hunger, not science: We're so good at noods because we were deprived of them for generations. Before the war, farming to survive. During the war, running to survive. After the war, rationed rice and dried sardines to survive. Then America—poor as dirt—still surviving.

So when we finally had money?

We went feral.

My people are notorious nood lovers. Some more notorious than others.

I ate like a king. I hit every market. Slurped every goddamn bowl. Burned my tongue raw. Blistered the roof of my mouth. Worth it. I gorged. I expanded. Gained what felt like ten pounds in one weekend.

And in between meals—now that I could walk like a semi-functional short Asian cyborg—I rode tiny electric scooters in endless loops around lakes and neighborhoods until my ankle staged a revolt.

That weekend gave me its first truth:

Healing isn't stillness.
It's motion.

Sometimes motion looks like a grown ass man on a child-sized scooter, burning his mouth and circling a Midwestern lake just to remember what being alive feels like.

Minneapolis feels like a living mixtape—each neighborhood its own genre. Tattoo shops. Coffee dens. Cracked asphalt. Uneven cobblestone. Earnest optimism. A city stuck in adolescence on purpose. Everyone's moving. Joggers. Strollers. Dogs. Maybe because winter's brutal here. You get outside while you can, or the cold swallows you whole.

It's a working-class cathedral of breweries, murals, Polish delis, corner bars, artists coaxing beauty out of cheap rent and bad lighting. It smells like yeast, spray paint, and hope.

Cross the Mighty Mississippi and it's college kids on scooters that sound like angry hair dryers, hauling pizza boxes like shields, convinced the universe is waiting for their debut.

You remember being like that.
You envy them a little.
You pity them more.

Minneapolis isn't postcard-beautiful. Not tropical. Not cinematic. Not easy. But it's beautiful the way real places are: imperfect, layered, scarred, stubborn, alive.

It's not a melting pot.

It's the only real salad bowl I've ever seen—distinct parts, messy edges, somehow working anyway.

And as you glide through it, you realize you're not just passing through. You're being absorbed. Even if only for a summer. A night. A ride.

I met strangers with stories heavy enough to crack your ribs. One man—didn't look homeless at all—asked me for seven dollars. Too specific to ignore. Said it was for a sandwich. I bought him one. And a drink.

He told me his wife gutted him. Took his kids. His mini coffee empire. Everything. "When there are four billion women on this planet," I asked, pretending not to know, "why let one ruin you?"

"Because it's easier this way," he said.

I understood that answer more than he'll ever know.

Minneapolis was a strange miracle. Proof my people can survive anywhere. Proof the King of Noods can too.

I came home lighter. Clearer. I gave my notice. They fired me a couple weeks later. The owner and his fake wife—with her fake body—gave me fake hugs and fake sadness, promised they'd pay what they owed.

Their eyes told the truth. They were lying.

Thank you for making this easier.

The next day I boarded an Amtrak west, heading toward Sacramento to pick up my truck from my dad. The company car was gone—the last thread tying me to that place—and now I was drifting through California, spine stitched together, mind loosening mile by mile.

Somewhere between the foothills and the climb, I pulled out my phone and wrote:

You're on a train winding slowly up a mountain. The world moves in patient frames—pine, shadow, rock, sky.

It's not all good.
But you're good.

Better than most.
Better than you ever admit, Cali.

Once upon a time, you would've killed to be this good.

Life is strange like that—unfair, ridiculous, unpredictable—yet on days like this, unbearably beautiful. The mountains stood ancient and indifferent, having watched every rise and fall and caring about none of it. Their silence made me feel small in the best way.

All around me: the hum of the engine, the rattle of tracks, steel whispering over steel.

That sound—the sound of a train climbing a mountain—has its own romance. For the broken. The exhausted. The ones stitching themselves back together.

Even in defeat, there's poetry.
Even in wreckage, dignity.
Even here, in my cold heart, a spark.

Just the quiet rattle of the tracks now, I wrote, *and a one-way ticket toward long-lasting peace.*

And for the first time in a long time—
I believed it.

Chapter 22: Sacramento

I hadn't expected much from Sacramento—a town of forgettable faces and shitty memories. The only thing it ever had going for it were the cheap, surprisingly good wineries on the outskirts—tiny oases drowning in boredom and money.

And yet there it was, five minutes from the train station, tucked on a one-way street beside a half-empty parking lot: a humble shop serving Lao khao soy.

Not the Thai version I tolerate but never crave—too curry-heavy, too loud—but the Lao kind. Usually polite. Delicate. Apologetic.

Except this one wasn't polite at all.

This bowl came out swinging. Not rich or flashy, but not afraid to make a mess either. Old world and new world thrown into the same ring, trading blows. Honest. Weird. Alive. It jabbed me in the gut like:

Yeah, I'm unexpected. Deal with it.

Two Lao women—owners, servers, mothers, everything—moved past me in skirts with more joy and authority than I'd unearthed in my last year of living. They floated through the space, smiling without effort, without permission. People who knew who they were. Knew where they came from. And didn't need approval to exist.

Sacramento. A city I'd written off—like a cheap cigar smoked to the nub and flicked into whatever gutter would take it. A place tied to old trauma, bad beginnings, and the worst version of me.

I always knew there were good noods hiding here. You just had to dig beneath the grime of Stockton Boulevard to find them. Historically, only my most desperate cravings ever made the pilgrimage.

Why bother? It was easier to drink and forget about her.

But this bowl—this honest, unpretentious bowl—grabbed me by the collar. Simple. Raw. Alive. A crack in concrete after winter. It made me question the way things were. The way things might be if I stopped carrying my ghosts like antiques.

Sacramento wasn't Minneapolis. Nothing is.

Minneapolis is number one in my personal rankings. Sacramento is maybe a distant tenth on a good day—lower than even Reno. I knew that. But suddenly it didn't feel like a graveyard. Maybe it still had a flicker of life. Maybe it wasn't a place that would kill me completely.

Because now it had things Reno didn't anymore—family. Food. Opportunity.

Reno was just ghosts and mediocre noods. A place I stayed out of duty, proximity, inertia—not because anything worth living for was still there. Except Group Therapy. And my girls would never try to cage me.

With no job tethering me, no relationship rooting me, no reason to pretend meaning still lived there—Sacramento didn't look like a downgrade anymore. If anything, it looked… responsible.

A city I'd dismissed, now quietly whispering:

You'd be a fool to bury me before I'm dead.

Then my old boss called—one of the few assholes who still held the title You've Never Beaten Me. Like vultures do when they smell change. But surprisingly, he wasn't circling to feed—he was offering a lifeline.

A real one.

A job in Sacramento. Doing exactly what he knows I do best.

Sales. Marketing.

Company car. Company card. The whole package.

Live rent-free with my brother. Work weekdays. Drive to Willows on weekends to help my parents. Be the good son again. Be the responsible man again.

And the math—Jesus Christ, the math was beautiful. Save nearly everything until the little one turns eighteen. Stack cash like a psychopath. Four hundred. Maybe five hundred thousand dollars. Enough to retire somewhere warm and quiet like a durian farm.

And when child support finally vanished into the ether—freedom. Rich and free.

Logistically, it was glorious.

A normal man would've said yes before the sentence ended. A smart man would've grabbed it and marched forward like a real adult.

But I've never been smart like that.

I've never learned to lie to myself convincingly.

I was already a terrible sleeper. Something inside me kept tugging. Whispering. Begging.

There has to be more.

More than fluorescent lights. More than Sacramento traffic. More than working all week just to drink yourself numb on the weekends. More than stacking money to someday buy peace.

Why not peace now? Why not life now?

I knew I needed to travel to heal.
I needed sun. Salt. Sand. Heat.
A hammock.
A bowl of noods that tasted like salvation.
I needed Thailand—or something like it.
I needed the ocean to talk to me again.

I'd promised myself the year before: Never make an important decision in Reno. Always make it on an island.

But my back was only weeks out of surgery. Fresh scar tissue. Nerves buzzing like exposed wire. A thirteen-hour flight felt impossible.

My spine flinched at the thought. My leg twitched. My body threatened revolt.

Still…

The islands were calling. Whispering through every breeze. Pulling through every gust. The way only destiny does.

I knew—even if I wasn't smart—I was close. Closer than I'd ever been. All I had to do was choose which path I had the courage to walk.

Mom drove the van down to my brother's place. Dad brought my Tacoma. The whole clan was waiting—my nieces, bright tornadoes of joy, and the demon baby boy who somehow learned to sprint before walking.

I played with them carefully. No bending. No lifting. No twisting. No BLTs. Especially around these feral creatures who treat your body like a jungle gym and a trampoline. I could feel the stitches humming under my skin like warning wires.

But God, it felt good to be alive around them.

I've always loved kids. Their smells. Their innocence. The way joy feels uncomplicated in their presence. The best years of my life I can never get back were when my boys were the same age.

I hugged everyone goodbye—sticky fingers, wild laughter, demon grin—and climbed into my truck.

The drive back to Reno felt like a test. Every bump a question. Every stop sign an evaluation.

My back answered slowly. Carefully.

But it answered.

When I got home, I didn't unpack. Didn't overthink. Didn't ask permission. I opened my iPad and booked a one-way ticket to Denver.

Why Denver? Because it was close. Because it was new. Because I'd never been. And because my back wasn't ready for anything farther yet.

Denver wasn't the destination.

It was the first step back into the world.

Chapter 23: Denver

Flying there was excruciating. My foot throbbed like it wanted to detach itself and take its own goddamn flight. Every minute in that cramped seat felt like an hour. But I made it. Barely. Breathing through spasms like a monk in purgatory.

Thank God Denver had scooters everywhere. They saved me. Saved my foot. Saved my sanity. I sought out noods first—obviously. Always noods. The compass of my soul points straight to broth and thin strands of salvation.

The next day I did what I always do in a new city—especially when my mind is messy, my back is fragile, and the world feels too big to take on in one bite: I booked a tour. A nice, easy, beautiful tour of Denver—the kind designed for old folks, curious wanderers, and lost souls trying to map their way back into life. The kind of tour that demands nothing except to sit, look, listen, and breathe.

I wanted history. I wanted bearings. I wanted those little morsels you can't get from Wikipedia—the quiet facts, the accidental lessons, the whispered stories a city only tells when you ask gently.

Morning light softened everything. Old brick buildings glowed. My back was stiff but manageable. My foot was still buzzing, but neither stopped me. I climbed onto the bus, settled in with a small wince, and let the city unfold: historic neighborhoods where Victorian houses leaned into the sun like aging socialites, murals exploding with color, graffiti that looked more like poetry than vandalism, parks still holding the cool breath of dawn.

The guide talked about gold rush fever, frontier dreams, railroad barons, miners buried under mountains, jazz scenes, prohibition tunnels, wildfires, rebirths. None of it will change my life. But all of it made the world feel bigger. Stranger. More possible.

I needed to remember the world is bigger than Reno—bigger than pain, bigger than injuries and corrupt bosses and tiny apartments and sleepless nights. Bigger than the version of life I'd been trapped in. I stared out the window and breathed slow, letting the landscape soothe me the way comic books used to when I was young.

Sometimes healing isn't grand. Sometimes it isn't a revelation or a dramatic moment. Sometimes it's just a quiet tour in a strange city. Tiny trivia. Random facts. A mind wandering like a child again. Sometimes it's just being somewhere else—and letting that place remind you life still has room to expand.

I scootered everywhere, farther and farther each day, testing the limits, testing the new back, testing my will, testing my hunger for life again. Denver became my training ground. My rehab dojo. My miniature freedom.

And then—one lazy afternoon—everything changed.

The sun was warm in that soft Colorado way, the breeze carrying just enough summer to feel like a promise. I had no plans. No expectations. Just riding my little scooter, breathing, drifting between shops, parks, alleyways, bowls of noods—trying to convince myself I'm not scared of the world anymore.

I sat at a Mexican restaurant bar. Football was on every TV. The shot glass clinked when it hit the wood. The reposado burned clean on the way down. The margarita dripped cold and sweet against my fingers. The bar A/C breathed on the sweat at the back of my neck like a blessing.

And I smell her first—warm, sweet perfume to my left.

Then I look up, and I see her. Jesus Christ. No—Victoria. Of course that's her name. A perfect name for a perfect woman. She isn't a type. She's her own category. A woman so beautiful and elegant she walks on her own red carpet. Everywhere she goes, men materialize out of thin air to buy her drinks. Exhausting, probably.

Denver looks like a gym exploded and scattered Patagonia- and Lululemon-clad fitness models everywhere. Everyone's jogging. Everyone's pretty. But her? She was beyond. From another planet. Well—Kazakhstan, which might as well be. Born there, moved to Kyiv, then Boston when her world blew up. She said it casually, like ordering a sandwich: war, bombs, relocation, mother, aunt, Airbnbs, books, shopping. Human horror condensed into a neat accent and perfect posture.

Maybe it was the height, the pouty lips, the long blonde hair, the piercing blue eyes, the perfect everything she couldn't see. That Eastern European exotic that makes men fever and women frown. I just wanted to sit in the same airspace for a minute, pretend I'm someone else, and try not to fuck this up.

She was at the bar flipping through Bourdain. My guy. One of my heroes. The only dumb shit I could come up with was: "How do you like the book?" She lit up. Said she loved how honest he was. Vulnerable. I told her I loved how much he loved people—and that it wrecked me when he died. We both nodded like we'd failed him personally. Robin Williams too. Mrs. Doubtfire himself. One of our favorite movies. One of our worst losses.

She told me I should be a travel vlogger. Said I have the personality for it—whatever that means. Probably a polite way of saying I talk too much. She was being kind. You're only kind if you've been through some shit. The beauty is just icing at that point. Kind is always better than smoking hot in my book. She had both.

Sheesh.

Then she told me where she's from, and I told her Dimash is my favorite singer and he's from there too. I offered her a ticket to New York to see him with me. She said her mom might know him, but she doesn't. I told her to bring the mom. She laughed.

She spoke Ukrainian, Russian, and damn near perfect English—with that dangerously sexy accent. A remote accountant, traveling like some nomadic goddess—one-way tickets and Airbnbs.

She doesn't eat carbs—unless they were tucked into a quesadilla at 1:30 p.m. on a lazy Saturday afternoon, stuck talking to a buffoon like me. She offered me a slice.

Jesus.

Generous too? Stop it, Man. Just stop.

She only ate chicken and fish. I asked if she refuses if it's been grilled next to other meat. She said no—it's not that deep. Not picky either. Somehow I liked her even more. How is that possible?

I travel like a man trying to outrun his own shadow—no plan except to find every nood shop in town. Cheap bed near all the action. Bottle at the end of it. She plans everything. Probably knows the menus and alleyways before she lands. A perfect partner: You plan. I show up. And try not to fuck it up.

At this point, all I could think was:

Don't fall in love, you idiot.

Don't you fucking dare. You're no good at it. You never were. Knowing that was the only thing protecting me from this kind of absolute rubbish. Love—it's bullshit.

For a moment I floated. I forgot my back.

Then she said she had to get ready for dinner with her mom and aunt. Of course. Plans. People like her keep plans. People like me run from them. She said she'd make plans for us another day. I said perfect, got her number, and tried to act normal. Inside I was fireworks on New Years.

We split in opposite directions. I walked away knowing I'd probably never hear from her again. I didn't think I'd be in her plans. And thank

God for that, I told myself. Because if she called—I'd pick up. And that… well, that would be the end of me.

I scurried back to my hotel like a lovesick bandit, heart pounding so loud I could hear it in my teeth. I couldn't believe what had just happened. Couldn't process it fast enough to file the moment away properly. I'd just spent a few minutes—mere breaths really—laughing with, talking to, standing in front of the most beautiful girl I've ever met in real life. Not online. Not imagined. Not an airport fantasy or beach mirage. Real. Breathing. Looking right back at me. Handing me her number like the universe was in a generous mood—for once.

I remembered the dudes at the bar staring daggers, drinks in hand, trying to figure out what cosmic glitch allowed me to pull that off. They hated the timing, the luck, the smile she gave me. They hated that none of them even came close. And probably hated most that they got beat out by some short, beat-up Asian guy. I didn't blame them. Hell—I didn't understand it either.

Walking back felt unreal, like the moment had too much glow around the edges to trust. I replayed the conversation, checking for cracks. Was that real? It couldn't be. It felt cooked up by jet lag, back pain, and too much tequila. What kind of day is this? Fly hundreds of miles with a half-healed spine, foot buzzing like a dying battery. Scooter around searching for noods and booze. Hop on a tour like a retired grandpa. And then—out of nowhere—step straight into some unbelievable moment?

Exhaustion hit me like a wave—physical, emotional, cosmic. I crashed on the bed staring at the ceiling, whispering: Did I just meet—and not completely fuck up—the most beautiful interaction of my life?

Calm down, Cali. Just a stupidly perfect girl who dropped into your orbit for one split second. Doesn't mean anything. I took a nap where my soul clocked out.

I woke up still floating. Scooted around to cool my nerves. Found a nood shop and slurped in silence, trying to prove to myself I wasn't dreaming. Then I grabbed some Jameson at a bar nearby to close out the night.

If this is what healing looks like, maybe I've still got a few surprises left in me.

The next morning, I sat down with my iPad—no plan, no intention to write anything monumental. I just wanted to breathe and jot down what I'd learned. But the second the screen lit up, the words poured out like they'd been waiting behind a door for permission.

I started writing about Denver. Not the city—what it did to me. What it cracked open. What it revealed. The sentences came smoother than anything I'd written in years—clear, honest, alive. I knew instantly: this wasn't random inspiration.

This was healing.

The back. The body. The mind. The heart. The goddamn soul. Pain is a package deal. I guess recovery is too. Sometimes the body heals first. Sometimes the mind finds light before anything else catches up. But as I wrote, something inside me softened—an old guard stepping aside.

Maybe I should stop fighting it. Stop trying to control every twist in the road. Stop treating healing like a checklist. Maybe I should let the universe steer for a change.

So I made a quiet vow—half joke, half hope—to not swear off women completely. At least not for a few days. Let the heart breathe. Let the story unfold without me wrestling the wheel away from fate.

But then there was her.

This sexy, brilliant, nomadic, one-way-ticket Ukrainian goddess who appeared out of nowhere—like a blonde hurricane of a muse. For her—for whatever strange thread tied that brief meeting together—for whatever destiny or destruction this was…

I surrendered.

To whatever this is. To wherever it leads. To whoever I might become because of it. For once, I wasn't chasing anything. Wasn't forcing a story. Wasn't trying to bend life to my will.

I became a leaf in the wind. I let the world take the first step. And I followed.

Denver

Denver. You Mile-High City, you. You beautiful, smug, oxygen-rich bastard. I want to hate you—truly I do—because you've got my heart in your hands and her lingering scent still on your breath. You've got my number and you know it.

You'd be perfect if you could figure out how to make me a decent bowl of noods. And lots of them. Jesus. It ain't that hard, Man. Boil the shit out of something meaty and long. Throw in noods. Add veggies if you must. Season with salt and MSG like a grown-up. Then feed it to me—consistently—exactly when I need it. How hard can it be?

You're what Reno wishes it could be if it kicked the meth habit, made the math work, shut down the strip clubs, slapped a curfew on the 24-hour booze circus, and stopped pretending degeneracy was some quirky personality trait. Don't get me wrong—Reno was pleasant for me once.

But you?

You don't wait around. You just do shit because it makes sense—saintly or cruel. No excuses. No speeches. Just action. They'll mention it in a history book real quick and keep moving. You walk the talk while Reno just struts in circles, flapping its mouth like I am right now—drunk on a Wednesday afternoon.

And the dogs. Jesus, the dogs. They outnumber the humans. I heard there are more dogs in Denver than kids. How is that even legal? People smile like they cracked some cosmic code—probably because they don't have kids. I get it.

Beautiful people only want more beautiful people, and in a sea of beautiful people, that's statistically impossible—so they get dogs instead.

Dogs don't cheat. Dogs don't ghost. Dogs don't demand you get hotter than you already are. Dogs love you even when you forget to text back.

Everywhere I look—scooters, joggers, bikers, shirtless fuckboys, sports-bra goddesses, running clubs, everyone hydrated to hell and back. I thought I was active once. Denver told me to shut the fuck up, sit the fuck down, put the Jameson away, and drink some electrolytes.

And the cleanliness—goddamn. My mom always said: If it stinks, don't touch it, Son.

But you?

I'd touch the hell out of you—just not in winter.

I dropped an AirPod on the sidewalk and didn't hesitate to pick it up. No piss. No mystery goo. No bugaboo. No bum sleeping on a pile of doo. A couple dog shits here and there, but hey—the dogs.

You even take care of your lost ones without turning it into a photo op or a moral Olympics. You hide your skeletons well. And those streets—brick and cobblestone—an old-world fever dream, a drunk construction guy's poetic wet fantasy.

Beaten half to death by your beauty—and hers—you softened me up. Soft enough to eat gelato in daylight for the first time in years. Soft enough to forget, for one fine second, that everything good eventually burns down.

You're too cool for school, Denver.

But thanks for letting me sit on your porch, sip twenty cocktails, slurp a few half-decent noods, spend a minute with my blonde bombshell soulmate—whom I'll probably never see again—and pretend I belong.

You're the coolest city I know.
Don't forget Ludlow…

Chapter 24: Seoul

She never called, but she did text back. That counts for something in this modern hellscape of half-ghosts and half-promises.

We never made good on our plans. She got sick—or had second thoughts—or maybe the universe didn't feel like lining things up that day. Who knows. But we stayed in touch. Long, thoughtful texts from two different worlds. Liking each other's Instagram stories like two ships flicking lanterns across the dark, yearning sea.

Probably for the best.

I got my fill of Denver anyway. More than enough. So I told her I was leaving. Mentioned I'd always wanted to see Korea—maybe—if my back held up.

She said she'd always wanted to see Korea too. Said it's beautiful. Said I should go, take pictures, and write about it for her. I don't think she had a goddamn clue how bad my back was, but a promise is a promise.

And I'm a man who keeps his word to a fault.

Deal, I said. She said deal too.

Then I flew home and stared at my healing spine like it was some delicate piece of machinery I wasn't sure belonged to me anymore. I wanted somewhere farther this time—farther than Denver, but not so far the flight would snap me in half.

Six hours sounded doable. Thirteen?

Not yet.

So I looked at Belize. Mexico. Somewhere warm. Somewhere cheap. Somewhere with water that didn't judge me. I researched. I paced. I caught up with friends. A couple offered me jobs. Another wanted to start a business. Bestie told me—with her usual brutal honesty—that

Reno had nothing left for me and I should be smart for once in my goddamn life and move to Sacramento.

But then I saw the prices. Belize? Mexico?

Those flights cost more than some monthly rents in Asia.

Something clicked—deep in the spine, deeper in the soul. I rubbed my back, whispered a prayer to the ancestors, took a breath...

And said: Fuck it.

I booked a one-way ticket to Seoul, Korea.

I was on the first flight out the next morning to San Francisco. Maybe the delay was my fault. It was late when I booked, and the options weren't exactly clear. And maybe—just maybe—I'd had one too many shots of Jameson.

A man gets bold after midnight and whiskey.

Either way, I spent sixteen goddamn hours at that airport. Landed in SFO at 7:30 a.m. Didn't leave until 11:45 p.m.

My back did not appreciate this terrible mistake, but the delay gave me Olympic-level time to walk laps between airport bars and ramen shops—hydrating and dehydrating myself at a perfect 1:1 ratio. I took shitty naps in those big yellow reclining chairs SFO loves to brag about. By the time the flight boarded, I had aged four emotional years.

Somehow, the long haul wasn't as bad as it could've been. I paid extra for the emergency-exit row—more legroom, less spinal death. Enough space to stand, stretch, and walk laps so my back didn't detach like dry-rotted fascia.

I didn't sleep a single minute.

Meanwhile, the cheerful dude next to me told me his whole plan—Seoul, then Busan—then fell asleep immediately for the entire goddamn flight.

I hated him.

I endured it with my usual pro-level traveler arsenal:

- Book about travel — check
- Neck pillow — check
- Backpack as emergency pillow — check
- Noise-canceling AirPods — check
- iPad loaded with forty hours of content — check
- Portable chargers like a seasoned fugitive — check
- Questionable life choices — double check

I landed in Seoul at 7 a.m. Exhausted. Delirious. Sore in places I didn't know still existed. The air was crisp—the kind that wakes your skin but not your brain.

But I was ready. Because I had my Denver windbreaker—the same one that carried me through scooters, sunsets, and the most beautiful disaster of a woman I'd ever met.

Groggy. Chilly. Half-dead. But there.

Korea.

The adventure was finally ready to begin.

I met a skinny white girl with short brunette hair while shuffling through immigration. Houston. Nervous but sharp. The kind of lost person who refuses to admit it.

Like me.

We teamed up out of survival instinct, trying to figure out the train. Two options: the expensive nonstop, or the cheaper one that stopped everywhere. Naturally, I took the nonstop.

Her hotel was nearby the center. She was in town for some Bitcoin convention—tech pilgrims and their temples. We failed miserably trying to hail taxis at first. Eventually we snagged a couple. Ladies first. We high-fived like backpackers in a movie scene and split.

My driver was talkative—half Korean, half broken English, the rest interpretive grunts. He loved basketball. I said I was from Las Vegas—easier than explaining Reno or my whole stupid backstory. The only American cities he knew were L.A., New York, San Francisco… and somehow San Antonio.

Why?

Victor Wembanyama.

He talked so much he missed my hotel and had to loop around.

It was only 8 a.m. Room wasn't ready. They took my bags—thank God—and I wandered like a sleep-deprived ghost through the streets, eye boogers and cotton mouth intact.

Gwangjang Market saved me—hand-pulled noods made by famous aunties. A bowl so fresh it practically grabbed me back.

When I returned, the room was ready early. Huge. Cheap. Seventy bucks. Aircon hummed to life.

I passed out.

Woke at 10 p.m. Hungry. Found Korean fried chicken—crispy, savory—washed it down with enough Soju until I was correctly drunk.

Survived the night.

Tomorrow would be for exploring.

The tour began with a Scottish guy pulling out a plastic bottle at a temple—thinking it was water.

Soju.

The guide nearly fainted. No trash cans. Public shame. Panic.

Naturally, I intervened.

"Allow me to take it off your hands, good sir."

He accused me of being an alcoholic. I denied it with dignity and took the blessing.

That's when I met Shin. My tour guide. Wiry. Smiling. A machine. I was the only solo traveler. Everyone else was old and moved slow. I glued myself to Shin.

First on, first off, never lost.

I walked more that day than I had in years. Feet splitting. Ankle swollen. Back screaming. Still proud that I even made it this far.

Pain is proof I'm alive.

Shin liked me. Asked to stay in touch. Said he'd show me real Seoul.

Then rush hour hit. Chaos. A human hurricane.

Never seen anything like it in my life and I lived in SoCal for twenty years.

I waited it out, limped into the subway, got lost once, cursed, tried again. Made it back.

Took a nap.

That night, Shin picked me up with his friend Chris—English tutor, dressed like Macy's America from head to toe. They hated Korea. Everything about it. Pressure cooker. No release valve. Beauty standards warped. Competition baked into every aspect of life.

We ended up at McDonald's—their favorite place in Korea.

Travel is funny like that.

Midnight came. Lights shut off. Shin drove me back through quiet streets. Before I got out, he said:

"You're a good person. You should really go to the Philippines, Brother."

Then he drove away.

And for the first time that night—

I believed him.

Seoul. Beautiful. Sad. Brutal. Brilliant. A city sprinting between past and future with no time to breathe. I was just passing through.

It left a mark.

But I knew—this wasn't the place I was looking for.

Not yet.

After a couple more days of blur—noods, Korean BBQ, soju, endless hop-ons and hop-offs, Gangnam photo ops, fancy libraries, fancier malls, bars, beers, shots, and enough photos to choke my camera roll—everything started to blend together. Beautiful, but repetitive. Same neon. Same crowds. Same skyline.

I needed a change of pace.

As much as I tried to be my usual happy-go-lucky self—striking up conversations, tossing smiles into the void—Seoul just wasn't having it. Everyone was too busy, too rushed, too buried in their own lives to talk to some wandering fool with a back held together by putty and Bondo.

All week it felt like I spoke to no one except Shin.

So I did the only thing that made sense: I booked a one-way ticket to Manila.

I got one last bowl from the always-open pho shop that never fails me, slurped it down like a final blessing, checked out, and hopped on the train to the airport—International Departures, where all wandering souls eventually end up.

I found a decent lounge to kill a couple hours. Two beers, a couple shots—enough to feel loose but not stupid. I stared out the big glass windows like I always do, watching other people's stories take off and land.

Boarded the plane with that perfect mix of calm and restless.

We pushed off the runway around 1 p.m.

Four and a half hours to Manila.

A short hop—just enough time for the body to settle, the mind to drift, and the next chapter to begin.

Chapter 25: Manila

From the plane, Manila doesn't sneak up on you—it sprawls at you. A massive, unfiltered, unapologetic metropolis, stretching in every direction like someone spilled an entire box of buildings across the earth and let it land however it wanted.

Endless from above. Stacked, layered, piled—concrete pressed against tin roofs pressed against high-rises pressed against shanties. New towers shoved right up beside crumbling ones. No pattern. No symmetry. Just density and poverty spreading wherever it damn well pleases.

There's a certain beauty in the chaos, but it's gritty. The kind that doesn't clean itself up for visitors. The kind that doesn't pretend. Pockets of neon. Clusters of malls the size of airports. And right beside them—rusted rooftops, tight alleys, laundry flapping like flags of survival.

Every block jam-packed. Every road alive.

From above, the city feels like it's breathing—loud, messy, unstoppable.

Manila isn't polished like Seoul. It isn't curated like Tokyo. It's raw. It's real. It's dirty in the way only a living, overworked megacity can be.

That's the sad charm. That's the naked truth.

Manila doesn't ease you in. It throws you straight into the deep end—no warm-up, no lifeguard, no chance to catch your breath.

Sink or swim, baby.

The first thing I noticed was the waiting: families packed behind rope barriers like it was the second coming. Mothers, uncles, cousins, crying babies—all squeezed together, waiting for someone who made it out and came back with stories. Or candy. Or foreign trinkets.

I stood there watching them and, for the first time in a long time, this cold prick cracked a little. Must be nice—having people who run to you the second you step off a plane. Who hold you and cry because love is leaving or coming back the only way it can at airports and hospitals.

They don't have much here. But they've got each other. In this kind of poverty, family isn't something you trot out once a year for Christmas photos—it's survival. A goddamn life raft in a country full of storms.

Then there was Estanick—my Grab driver.

A complete lunatic. A kamikaze in flip-flops.

Texting. Swerving. Dodging motorcycles like it's a video game set to hell mode. Casually talking to his mom on speakerphone while racking up twenty near-death experiences in under an hour.

His horn wasn't a warning.

It was a declaration of war.

I should've taken a shit at the airport. I'm pretty sure I left something spiritual in his backseat. The way he cut off four rows of traffic downtown—in the rain, during rush hour, while making a full ass U-turn—was absolutely wild.

The craziest driver I've ever seen in my whole goddamn life.

Puts Yan to shame.

And yet the madman got me to my hotel in one piece.

Welcome to Manila, motherfucker.

It was already dark and lightly raining when I checked into my hotel—quaint, clean, staffed by people so warm it felt like they'd been waiting for me their whole lives.

I grabbed a quick bite at the rooftop: fried pork belly, a runny egg, garlic rice—the holy trinity of Filipino soul food.

Washed it down with a cucumber cocktail and watched the city flicker beneath me. Beeping. Brakes. Sirens.

The night was calling.

The closest bar was only a couple blocks away, tucked into the rowdy heartbeat of the red-light district. Early still—two old white guys at a table, eight or ten girls lounging around, Mamasan scanning the room with military focus.

I walked in like I owned the lease.

No bell to ring, so I sat at the bar and immediately yelled at the bartender to pour a round of shots for everyone in the goddamn place—punctuating it with a swirl of my finger in the air, international sign language for:

Let's fucking go!

They love me in Thailand, but here?

Jesus Christ.

They loved me even more. Taller, louder, funnier—and every one of them spoke English. Easy conversations. Easy laughter. Nothing forced.

Sure, they could smell money on me before I even sat down. Everybody in those streets can.

But I wasn't there for cheap sex. Not my style.

I was there to sing—loudly, terribly, religiously.

Because everyone knows Filipinos are the greatest singers in the world. Not the off-key howling you get back home. No—these drunk voices could win American Idol while smoking a cigarette, then go cook you dinner.

I told the DJ, "Play Akin Ka Na Lang by Morissette—my girl, the greatest female voice alive."

He blasted it, and instantly every girl swarmed me like I was front row at a concert. We danced, shouted, butchered the whistle notes, broke down laughing, and demanded a repeat.

The DJ understood his assignment.

I bought the cheapest bottle of watered-down tequila they had—economical debauchery. The girls kept trying to drag me to my hotel—playful, persistent, professional.

I told them no.

Too easy. Too gross. Too meaningless.

They respected it.

They still tried.

I took off early, to their dismay. I think I gave them more fun than they'd had in a while without touching a single soul.

They hugged me, fake-pouted, grabbed my phone, swapped WhatsApps, promised to see me again.

I told them maybe.

Which, of course, meant yes.

Tomorrow would be for exploring.

And noods.

Always noods.

Manila wakes up loud.

I opened my curtains to a sky thick and gray—heavy, suspicious. Breakfast was another greasy, hearty rooftop plate: tocino, garlic rice, egg—washed down with instant coffee that tasted like burnt determination.

I headed to 7-Eleven to find pesos for laundry, dodging dangerously low power lines that were definitely not OSHA-approved.

The streets were already alive: barefoot kids weaving between jeepneys held together by prayer, vendors grilling meat over barrels, old women selling mangoes and boiled eggs in plastic bags.

Everything sticky, chaotic, and somehow functional.

Everyone was nice. Everyone spoke English. Everyone tried to help. This country is a drunken boxer—dirty, sweaty, ribs broken—still smiling, still dancing, still punching way above its weight.

They've got nothing, and somehow they still try to give you everything.

I ducked into 7-Eleven for two minutes. They didn't have change. I turned to walk out—

—and stepped straight into a catastrophe.

A full-blown typhoon landed in the span of a breath. Water rushing ankle-deep. People sprinting, laughing, screaming. Tarps flapping like broken wings. Vendors dragging plastic sheets over stalls. Jeepneys flashing hazards. Motorbikes diving under awnings.

I stepped outside and got instantly soaked—like someone dumped a bucket of the Pacific Ocean over my head. I had to turn right back around and buy a cheap umbrella to survive the walk back.

Eventually the sky calmed its tantrum. Streets steamed. I stepped back out, dropped off my soaked clothes at a newly opened laundromat, and found a humble family-run restaurant nearby.

The old man running it had the softest smile.

I ordered two San Miguels and did research on where to go next.

My research was YouTube.

The first video sold me instantly:

Coron, up in northern Palawan—less crowded than El Nido, less storm drama, more raw beauty.

I booked the flight and hotels on the spot.

The plan appeared fully formed: start in Coron, wait out the storms, head to El Nido after they pass, then hop islands until my back healed, my spirit reset, and my physical therapy got completed on a beach warm enough to forgive all my sins.

After that, I headed to Binondo—Manila's Chinatown, the oldest in the world.

And it looked old.

Not romantic old.

Exhausted old.

Dirty, cramped, choking with traffic. The driver warned me to hurry—when the typhoon hits, Chinatown becomes a flooded, muddy island with no way in or out.

I thanked him, wandered a few streets, found a shop, ordered pork noodles—too heavy, too porky, too much of everything.

Ate three bites and surrendered.

Grabbed a ride back before darkness swallowed the city. Picked up my laundry, packed it neatly, and went back to the rooftop bar with two more San Miguels to stare down at the chaos I'd somehow grown fond of.

And of course, I ended the night exactly how I ended the one before:

back at the red-light bar with my new favorite girls, screaming Akin Ka Na Lang until my voice gave out and the bartenders begged us to shut up.

Crawled back to my hotel, half-drunk, half-deaf, fully alive.

Chapter 26: Coron

The next morning I grabbed a taxi to Domestic Terminal 3, checked in, slipped through security, and immediately felt the heat of the place—the kind of thick, sticky Manila humidity that hits you like you've just stepped into someone's armpit. But once everyone settled into their waiting seats, the giant industrial fans started pushing enough air to make it bearable.

I grabbed a couple empanadas for lunch and waited for the inevitable:

Delay.

These old domestic prop planes haven't seen an upgrade since the Marcos era. A little mist, a little rain—maybe a bird sneezes somewhere—and it's delays and cancellations across the board. Ours got pushed back an hour. Standard stuff, as I'd find out shortly.

We finally took off around 3:30 p.m., and the short one-hour flight ended with a descent that punched a lump straight into my throat—electric turquoise water bleeding into endless jungle emerald. Even through mist and cloud cover, it was the most beautiful approach I'd ever seen. Something inside me stirred—some ancient recognition, like my bones knew this place long before I did.

We landed on the northern flat stretch of the island and I hopped into a shuttle with a quiet Filipino family heading south into Coron Town. The ride wound through hills, lush greenery, shacks, dogs lazily crossing the road, and open palms until the port slowly came into view.

I checked into my ideal room—clean, simple, small bathroom with a bum gun, the standard "insert key to make the electricity work" slot on the wall. Dianna's Inn. Twenty dollars a night. Absolute robbery on my part.

The place was about five blocks northeast of the town center—an easy walk, but I didn't feel like walking. I tossed my bags down, opened Google Maps, found a chicken joint, and stepped outside.

A tricycle driver spotted me instantly, waving like he'd just seen his long-lost cousin. I climbed in. The ride cost 50 pesos. Less than a dollar. I gave him 100.

The man lit up like Christmas came early.

Pure joy. Pure gratitude. Pure Philippines.

I filled my belly with BBQ chicken and wandered through Coron—a port town so small it threw up four streets, shrugged, and disappeared back into the jungle.

I needed some degeneracy to unwind from the day's traffic, noise, and delays, so I walked into the first place with a neon sign: **Big Kahuna Bar & Restaurant.**

Empty as a church on Monday.

Just me and my ghosts at 8:30 p.m.

Perfect.

And that's when I met Zoey.

The bartender.

Mid-length brown hair brushing her slender shoulders—probably why she wears those wide, loose tops, so the collarbone can do what collarbones do: quietly steal the show.

Big eyes. A cute button nose. Full, pouty lips. And the lightest skin I'd seen on this island—shade that suggests she avoids anything outdoorsy... or maybe avoids people altogether.

Which, honestly, might be why she's so damn pretty.

I ordered the first cocktail on the menu—a Hurricane, fitting since a typhoon had just passed—and a double Jameson because habit dies harder than hope.

The bottle was on the top shelf, way out of reach for her. She dragged over a big wooden chair and climbed up like a kid reaching for candy. Couldn't have been taller than four-ten, but she didn't wobble. Completely sure of herself.

I used to be like that. When the world still felt like something I could bend.

Right away I could tell she wasn't like the others. Too sharp. Too something I couldn't name. She said she was old enough to pour booze, though she barely looked it. People here stay eighteen until they're eighty. Asians don't raisin.

What caught me was her accent. My Phuket guide had that thick British-Thai mash, but Zoey's flipped it—almost-too-polished British, with a ghost of Filipino beneath it, so soft you'd never notice unless you leaned in.

"English major," she said. "One year left."

She'd picked up her accent watching *Game of Thrones*, *Bridgerton*, Marvel movies. Netflix University. YouTube Seminary. Brilliance pretending to be laziness. I respect the hell out of that.

She said she could do other accents too—a bit of French, even—but didn't really know what an American accent was.

"Easy," I told her. "Strip away anything good about the English language and add a shit-ton of swearing and sarcasm."

I demonstrated.

She was "fuckin this" and "fuckin that" in five minutes.

Proud motherfucker right here.

She told me she was single because boys here were stupid.

"They're happy rotting where they stand," she said. "Just cheap alcohol at the corner, half an education, no dreams."

No ambition.
No clawing their way out.

"They've made peace with the ceiling," she said, "and I never will."

God, I hoped she never settles.

Then she asked what I was doing here.

"Tired of Korea," I told her. "Got bored last week, stumbled down here looking for decent people after getting ignored all over Seoul."

Her eyes lit up—Korea was one of her dreams.

And there it was.

The missing piece. The spark I felt earlier.

She's saving pesos to get the hell off this island—Korea, England, France, anywhere bigger than this puddle she was born into. I showed her photos of Seoul, and a few things I'd written. She tore through them like someone starving for something beyond herself—asked questions that weren't cheap, the kind of questions you only ask if you're already half out the door.

She wanted to meet Shin.

Maybe I'll play matchmaker—he needs someone real, and she needs someone who understands what it's like to be a stranger in a strange land.

She's outgrown this place. Petite girl, massive teeth-cutting dream. Been dreaming since she was small, she said. Always felt something was waiting out there, and she'd wrestle it into shape eventually.

She reminded me of me at eighteen—moving to L.A. broke, dumb, and drunk on dreams. Thought they'd carry me to the promised land. And they did, until they didn't. I burned through all of them like cigarettes—some tragic, some hilarious in their ruin—and now I wander countries chasing the next bowl of spicy nood like my life depended on it.

And maybe it did.

But her?

She's different.

Too bright to dim. Too stubborn to sink. Too hungry to stay here and rot.

She doesn't need luck. She'll take whatever she wants and wring its neck.

She'll be more than fine.

I made her pour me another shot—tipped her a hundred percent toward her Korea savings account—and took a tricycle back to my hotel.

I woke up the next morning more refreshed than I'd felt in a long time. Maybe it was the island air finally cracking something open. Maybe it was the promise of snorkeling among the impossibly beautiful islands I'd been staring at on screens for months. Maybe my back had finally loosened its grip. Maybe it was because I finally had a real conversation with an actual human being last night—my first one in over a damn week—that something inside me unclenched.

Whatever the reason, I wasn't going to overthink it.

I wanted a cheap, simple breakfast—the kind that requires zero thought and delivers absolute satisfaction. Two blocks down, I found exactly that.

I crossed the already buzzing main road—tricycles clanking, engines choking and coughing, horns announcing their existence like morning roosters—and sat at a low plastic table on a low plastic chair at this open kitchen of a restaurant by the side of a busy dirty road. My type of restaurant.

I ordered a simple beef soup with cabbage, a bowl of rice, and a tub of hot sauce that could strip five layers of paint. Ate it sweating and smiling, wiping my forehead with napkins so thin they dissolved if you wiped more than once. The only coffee they had was instant, like most places on this island.

Between bites, I watched the circus of transportation unfold in front of me.

Tricycles—thousands—swarmed the streets. No buses. No trains. Just motorcycles welded to tin sidecars with bald tires, bent axles, and paint jobs that looked like a fever dream. Chrome, saints, stickers, rust—each one a Frankenstein masterpiece built by a man who's been holding a wrench since birth. You're not a man until you build your own steel chariot.

They hauled everything—tourists, families, groceries, propane tanks, school kids, drunks like me—the entire ecosystem of a town living on four square streets, moving in perpetual chaos.

And the dogs. Jesus, the dogs.

They navigated this metal stampede with supernatural sense and grace. No leashes, no owners shouting commands—just street-smart creatures weaving between tires like they'd passed some secret canine road exam.

I sat there, sweating, slurping beef soup in the humid morning heat, watching this tiny town run on rust, fumes, and determination.

There's something oddly beautiful about a place that doesn't need perfect roads or perfect systems to function.

Just people who refuse to stop moving, stop living, stop being content.

It made me envious.

Breakfast done, spine surprisingly quiet, and heart weirdly light, I stood up, paid my tab, and headed off toward the port—ready to let the islands punch me in the face with whatever magic they had waiting.

And let me tell you—Coron beat the living shit out of me in the best possible way.

From the moment I stepped onto that rickety plastic bubble-wrap of a dock, I knew I was outmatched. Even the boats looked smug—long, spindly bangkas with bamboo outriggers like dragonfly wings, painted in tropical pastels as if they were auditioning for an epic movie.

I met Erik and his wife from the Netherlands—kind, sunburned travelers who just happened to choose the wrong seats and ended up with me instead of peace and silence. They told me they'd spent the last five days trying to outrun typhoons tearing up from the south of El Nido. Terrifying for her, exhilarating for him. He said it with a grin you only earn by cheating death. They even dodged a few pirates, he bragged. Now they were here, hoping Coron would give them more beauty than the scattered days they salvaged before the storms came.

Chasing serenity after surviving chaos—fair trade, if you ask me.

We motored out into water so aggressively turquoise it looked unreal— Photoshopped by a god who doesn't understand subtlety. The sun hit it and the whole ocean lit up like neon—almost rude in its beauty. The kind of color that makes you question whether you've ever seen blue before, or if every other body of water was just a shitty knockoff.

I sat there on that wooden bench, life vest half buckled, back twinging, holding onto the boat with one hand and my dignity with the other, thinking:

This is what people risk their lives for, like Erik and his wife,
and this is why fools like me cross the planet with stitches in my spine, chasing life.

Because this kind of beauty demands courage and pilgrimage, not cowardice and permission.

The first stop was a lagoon that required stairs carved out of jagged limestone. Everyone seemed to be sprinting up them like baby goats. I was the last one as always. Ten steps up and I was already exhausted. I lumbered behind them, sweating like a sinner from last night's booze, heart pounding, pretending my spine wasn't about to stage a tragic protest.

Only two hundred ninety more to go.

You got this, Cali.

At the top, a narrow walkway curved along a cliffside, opening to a hidden lake so clear and deeply turquoise it felt divine—like someone drained heaven and poured it into a magical crater.

I jumped in and the weird mix of cool and warm water hit me like a baptism and a slap at the same time.

Then—

Back pain? Gone.
Anxiety? Gone.
Self-loathing? Gone.

For a moment, I wasn't a man recovering from surgery, drifting through countries, searching for meaning in noods and bars.

I was a stupid speck of dust floating in ancient water, staring up at limestone spires that looked like broken teeth chewing up half the sky.

Then it was time to go again—like all tours.

We climbed back up and down, boarded again, and Coron kept swinging.

Island after island, lagoon after lagoon—every stop more ridiculous than the last. Wooden huts with woven roofs perched on white sand so fine it squeaked under your feet. Grandmas selling coconuts—chopping with machetes longer than their arms. Dogs napping in the shade like they owned the beach. Signs that say: **Danger – don't pet cat.** Wild chickens wandered the edges of our paths, pecking lazily at the day. Kids swimming in makeshift Styrofoam canoes held together by string and wire, selling you handmade trinkets.

And the fish. Goodness gracious, the fish.

I've fed the fishes with my puke in Maui. I swear these were even more beautiful.

Neon parrotfish, angelfish, little black-and-white ones swarming like tiny mafia soldiers whenever you dropped rice. I snorkeled above them like a bald, uncoordinated sea god—kicking awkwardly, trying not to drown, silently praying my spine wouldn't detach mid-breath.

They ignored me completely—like I was just another dumb floating tourist cluttering their reef, trying to take videos with the stupidly expensive GoPro he bought because he fell in love with a blonde goddess.

Rightly so.

Between stops, the boat cut across open water—wind perfect, salt spraying, everyone sunburned, silent, stunned—united by the rare experience of collectively shutting the fuck up and just staring at unbelievable sights.

Coron will do that to you.

Not because it's peaceful.

Because it's overwhelming. Especially to the uninitiated.

Beauty as violence. Wonder as assault. Nature as an uppercut.

By sunset, between the sputtering engine of our bangka, the sky bled orange and studio mauve, clouds hanging low like lazy gods, and the water mirrored it so perfectly it felt staged. I stared at it with the kind of dumb gratitude that makes your chest hurt.

I thought about all the places I'd been chasing peace, all the bullshit I'd dragged around with me, all the miles and surgeries and heartbreaks and noods—and realized the joke:

I didn't have to fight for this moment.

Like most things in life, I just had to show up.

Coron would do the rest.

It didn't ask me what I wanted, or what I carried, or what I was trying to patch, or what I was running from. It just opened its arms and beat me senseless with its beauty.

Every dock, every cavern, every absurd shade of sapphire, every shaky stairway cut into sharp rock, every tattered hut, every fish, every sunset—bruised me in ways I didn't know needed bruising.

I made it back to shore sunburned, salt-crusted, exhausted, and borderline euphoric. A broken man with a full heart, limping down a wet dock in flip-flops, whispering to himself:

This is more than I deserved.

And that was Coron's kindness.

Not softness.
Not comfort.

A beautiful beating—administered with love.

It was almost dark already when I grabbed a tricycle back to the hotel—wind in my face, spine complaining, island dust settling into every fold of my shirt. Quick shower, lots of lotion on my extra crispy fried face, clean clothes, then the familiar walk down to my favorite bar to see my favorite bartender.

Chapter 27: Brick to the Face

Turns out she wasn't four-ten after all. Five-two, she says. According to her, that's tall. Sure, babe. A skyscraper in a sea of tan dolls. I buy it.

She was out of Jameson. How the hell does a bar run out of Jameson? Oh yeah—I was here last night. My fault. So I made her climb the goddamn chair again—little queen on her sturdy throne—and grab the Tullamore Dew instead. That'll do. And another Hurricane too.

I slid my scribbles across the bar. Told her it was half-finished but worth reading. Told her it was about her. She hesitated—some polite excuse about not wanting to read something incomplete. A flicker. Quick. Gone. Interesting. I insisted.

She finally read it—laughing, sighing, embarrassed. Then serious. Said it was all true. Said it was strange seeing yourself through someone else's eyes. Victoria said that too. Maybe for once my greatest weakness—seeing only the good in people, not the shadows they hide—had turned into a kind of strength.

I told her I'd introduce her to my buddy Shin. Maybe set them up.

The iron little empress melted. Fixed her hair. Dabbed lip gloss. Covered her mouth. Cheeks on fire. There it was. A crack in the armor. For a split second, she wasn't the razor-sharp island bartender with a polished accent and an even sharper bullshit filter. She was just a twenty-three-year-old girl—soft, bright, human.

So I called the coward. No answer. Texted back asking for pictures first. Then said he was going to nap. Jesus Christ. Men say they want queens. Then hide when one shows up.

She snapped right back into her fortress—chin lifted, eyes steady, voice clean.

Born in the deep south of Palawan—poverty and dirt as a starting line. Saw too much too young. Knew she'd get out the moment she could. Knew before she could spell. Knew English was her ticket out of hell. An aunt married north, drifted to Coron—slow days, quieter nights. She followed after high school. A roof. Food. A university nearby. A move from chaos into clarity.

Smart. Focused. Efficient. Knows her worth. Knows the difference between boys and men. Between leeches and friends. Between settling and reaching.

Parents? She doesn't go there. My questions faded. She let them hang, practiced and precise. Her last name didn't match the one online—her mother's. Worn without explanation. No apology. No bitterness. Silence explained. Wounds implied.

She didn't flood the internet with posed perfection. No curated body, no thirst traps, no borrowed confidence. Good. A rare kind of wholesome. A calm center in a loud world—not selling herself, not hiding either, just standing exactly where she is.

Then she grabbed my phone, took a hundred selfies, deleted every one.

"Hey—my phone! My pictures!" I protested. She said my phone was the problem. My property, sure—but in her world I was just visiting.

I told her it felt like everything had been building toward something. An itch I couldn't scratch. Flying far. Running from scars. Meeting strangers in strange bars. Victoria in Denver—told me to travel, to write. Shin in Korea—told me the good ones were south. Then her—in a random island bar—and I try to hand her off to someone too scared to reach for the stars.

She said it casually, like she was commenting on the weather.

"You just need to dream bigger."

And it hit me like a brick to the face.

After I caught my breath—after her small, knowing smirk—I said, half stunned, half surrendering: "Maybe it wasn't about Denver. Or Korea. Or the delays and detours. Maybe it was about me meeting you. A broken dreamer from the refugee camps crossing paths with a kid from the slums whose dreams are still intact. That's why I'm here. Right now. At this exact moment. In this place. Out of all the vastness of time and space—so you could look me in the eye and tell me I wasn't dreaming big enough."

She tilted her head. Smiled.

"That's fuckin' deep," she said.

And just like that, the itch was gone.

Son of a bitch.

Chapter 28: Leaving Paradise

They call Thailand the Land of Smiles. But the Philippines? This is the Land of Kindness. They don't ask. They don't expect. They just give.

After saying goodnight to Zoey, I tried to bum a cigarette from a group of tricycle drivers at the corner. None of them had one. They buy cigarettes one at a time here—packs are a luxury. That surprised me. I bought a full pack and passed it around. You'd think I'd handed out gold bars. One of them—Sunny—offered me a ride back to my hotel. I left the pack on the seat, figured it was a gift.

Two minutes later, there was a knock on my door. Hotel staff. Holding the cigarettes. "Your driver said you forgot these." I was speechless.

In Thailand they smile. Here, they return what you didn't know you lost.

The next day I gave Sunny a thousand pesos—twenty bucks to me, probably rent to him. He nearly cried. Now I've got brothers at the corner bar. Free rides. Numbers in my phone. Brothers from other mothers.

I went to my favorite breakfast spot—hearty beef noods, cheap and perfect. I've always believed food reflects people. If it's soulful, the people can dance. If it's bland, it's European. If it's spicy, it's Mexican or Asian. This food was simple, soulful, and way too good for the price. Pure Filipino.

Then I took a tricycle to the pier for another island-hopping tour. No adrenaline. No rushing. Just three impossibly beautiful islands and a freshly cooked lunch in between. While everyone lounged in the shade, I turned the tour into physical therapy. Slow walks up white sand beaches. Step into the water. Float. Let the ocean lift the weight off my spine. Walk again. Rinse. Repeat. Everyone else was here for selfies. I was getting burned alive and stitched together at the same time.

Last night fed my soul. Today fed my body.

Eventually, I found shade—on a lazy swing between two coconut trees—and wrote:

I was lost, but lost isn't always missing.

And there You are, God, pulling strings from behind the curtain—sneaking up on me through a tiny, feisty Filipina who told me I wasn't dreaming big enough. Well played, You cosmic bastard. I get it now. You took everything from me so I could see. You can't fuck up something that was meant to be. You want big dreams? Big fucking dreams you shall have.

That night I told Zoey the truth—that I was a selfish wandering drunk, stumbling across continents looking for something to believe in. I didn't know what I was searching for. Just a pulse. A hunch. But I figured if I traveled far enough, wandered long enough, and slurped enough goddamn noods, I'd eventually trip over it.

First came peace—on a hammock in Koh Phangan last year. Then came health—under a knife not long ago. Then came the last piece I didn't know I needed.

I found it.

My dream.

She looked at me sideways—half skeptical, half soft. "What's your dream?" she asked, not expecting anything real. So I showed her what I'd typed the night before:

To dream big so I can give back to the poor.

She didn't clap. Didn't laugh. Just nodded. Some things don't need applause.

I thanked her properly. Told her I'd be rooting for her from afar. She barely flinched. That familiar smirk said everything: I don't need anyone rooting for me. God, I hate her. I tipped her—double. Hugged her. Told her I had to go. That big dreams were waiting.

What she didn't know was that the dream arrived holding the hand of a seven-year-old kid. It was always a package deal. Lose yourself, lose the dream. But when the dream comes home—that kid does too.

The next morning I called my future partner. No easing into it. "I'm coming home, Brother. You're building me an office. You're giving me half your company. We're gonna be broke for a while. But we're gonna build something big—not for the sake of money. We're gonna give back. We're gonna dream big."

Silence.

Then: "Oh… fuck."

Followed by: "Fuck yeah! Let's go!"

The first brick was laid.

I packed up, tipped the cleaners, bowed to the smiling front desk girls. They bowed back like they'd known me forever. Then I crammed into a shuttle with a loud Russian arguing over fifty pesos like it was a hostage negotiation. It started drizzling. Delays stacked. Then: cancelled. Anywhere else, riots. Here? Shrugs.

Paradise breeds patience.

I ended up back at the hotel. Tried to surprise Zoey. The typhoon said no. I ate sisig again, drank a San Miguel, and went to sleep.

Morning. Delay again. Missed my flight. Booked another. Sat sweaty and broke in Terminal 1—smiling.

Called my mom. "I'm safe, Mom. I love you. Your son is finally coming home." Then the truth: "But I'm not coming back to Sacramento. I'm going to Reno. Let me try, okay? Let me fail on my own terms."

She paused. "Took you almost three years to heal," she said.

"I'm slow."

She laughed.

"You're fine now," she said softly, knowing she won't need to call me every day anymore. One day I'm going to miss that. We hung up.

On the plane home, staring into the black window, my own face floated back at me. They say travelers are either running from something or searching for something.

Turns out, what I was searching for was waiting in my childhood—same humidity, same sounds, same kind of people. Not a place. An orbit.

And then it hit.

The dream.
Bright.
Clean.
Unmistakable.

And for the second time in my life—
I was ready to look it in the eyes.

Epilogue:

The Big Dream

"You don't want to go to America, boy! White demons will pick you up with chopsticks, stir-fry you, and wash you down with wine!"

That's what my grandparents told me in the refugee camp.

Chieng Kham. Tin roofs. Dirt floors. Barbed wire. Helicopters dropping dried sardine and rice rations like mercy once a month. We were the ghosts of a war nobody wanted to remember. Nomadic phantoms cursed from the beginning of creation to suffer eternally, stuck between jungles and borders, tucked halfway between nowhere and despair.

The elders wanted to stay. Familiar dirt beats unknown promises.

But my parents had fire.

America or bust.

And me?

I was a seven-year-old kid with a dream the size of a Pepsi bottle.

The first time I tasted Pepsi, I thought it was nectar from the heavens.

Men pooled their life savings and threw a party—Pepsi, Sprite, 7-Up. The women weren't allowed. The kids weren't either.

I followed my dad like a shadow as always. Dad told me to go home. My favorite uncle laughed and handed me a cup.

Fizz and lightning.

The world cracked open instantly.

That night, staring through the barbed wire fence, I promised myself:

One day I'll go where this drink comes from.
One day I'll come back a king.
And I'll give it away to all these poor kids.

That was the dream.

I buried it. Outgrew it. Smothered it.

Until it found me again—on an island, in a bar, across from the perfect messenger who told me I wasn't dreaming big enough.

Turns out…

It was never gone.

The End.

Actually—

Just the beginning.

Kao "B.O.L.O." Lee is an author and builder whose work explores identity, reinvention, and the strange beauty of starting over.

Born in Laos in 1979, he spent his early childhood in refugee camps in Thailand before immigrating to the United States with his family at the age of eight. Raised in Northern California, he became the first Hmong valedictorian of his high school and later earned a full-ride scholarship to California State University, Long Beach, where he completed his undergraduate studies.

In 2018, Lee published his first major book, ***All In: Beat the Odds in Sales.*** He also writes children's books dedicated to preserving Hmong folktales and cultural stories for future generations.

Today, he owns and operates a general contracting business in Reno, Nevada—built on the mission of giving back to the community and serving those in need.

The Drunk

Made in the USA
Coppell, TX
07 February 2026

70579878R00105